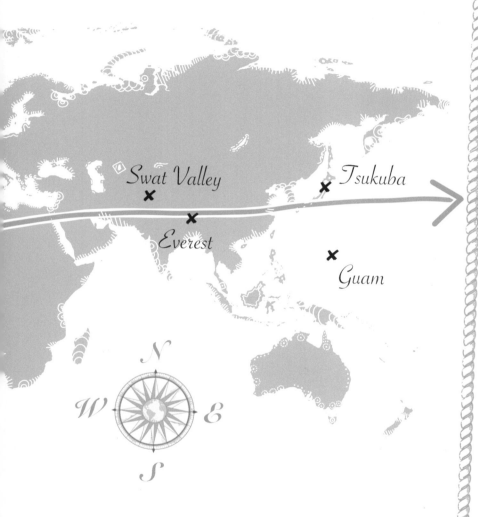

Swat Valley

Tsukuba

Everest

Guam

N

W   E

S

JET STREAM

Jet Stream

TIM WOOLLINGS

*A Journey
Through our
Changing
Climate*

OXFORD
UNIVERSITY PRESS

OXFORD
UNIVERSITY PRESS

Great Clarendon Street, Oxford, OX2 6DP,
United Kingdom

Oxford University Press is a department of the University of Oxford.
It furthers the University's objective of excellence in research, scholarship,
and education by publishing worldwide. Oxford is a registered trade mark of
Oxford University Press in the UK and in certain other countries

First Edition published in 2020

Impression: 1

Published in the United States of America by Oxford University Press
198 Madison Avenue, New York, NY 10016, United States of America

British Library Cataloguing in Publication Data
Data available

Library of Congress Control Number: 2019937512

ISBN 978–0–19–882851–8

DOI: 10.1093/oso/9780198828518.001.0001

Printed and bound by
CPI Group (UK) Ltd, Croydon, CR0 4YY

To Ella and Clara

ACKNOWLEDGEMENTS

First and foremost I'd like to thank my wonderful wife Helen for all her help and support during the writing of this book. I couldn't have done it without you, not least for all your careful editing, guidance and of course patience! Many thanks are also due to the rest of my family for all your support while I've been busy writing. Special thanks also go to Claire Delsol for her hard work and artistic skills in making many of the figures for this book including the brilliant map.

This book would also not have happened without the help of Ania Wronski, my editor at OUP. Thank you, Ania, for all your help and advice and for being interested in this project in the first place. I am also very grateful for the many constructive comments from anonymous reviewers of the proposal for this book.

I am deeply indebted to several people for providing invaluable comments on all or part of draft versions of this book. The following people gave comments which were very gratefully received and have helped to significantly improve the book: Peter Watson, Robert McSweeney, Sven Titz, David Pyle, Cheikh Mbengue, Liping Ma, Giuseppe Zappa, Maggie Watson, Olivia Martius, Nathalie Schaller and Mio Matsueda.

This book has also benefitted from many useful discussions with lots of people who have taught me new things about the jet stream, meteorology and its history, so thank you in particular to Brian Hoskins, Camille Li, Hans Volkert, Christoph Raible, Libby Barnes, Peter Read, Adam Scaife, Rodrigo Caballero, Geoff Vallis, Peter Rhines, Clara Deser, David Stephenson, Andrew Charlton-Perez, Tim Palmer, Rowan Sutton, Mike Wallace and Henrietta Harrison.

I'd also like to thank Tess Parker, Chris O'Reilly, Hugh Baker, Ronald Li, Marie Drouard, Matt Patterson, Lynn McMurdie, Lesley Gray, Noboru Nakamura, Ed Hawkins and Giles Harrison for their support during this project and Mike Blackburn, David Brayshaw, Richard Greatbatch, Hisashi Nakamura, John Methven, Ben Harvey, Len Shaffrey, Ted Shepherd, Helen Dacre, Robert Lee, Abdel Hannachi, Joaquim Pinto, Giacomo Masato, Kevin Hodges, Christian Franzke, Keith Williams, Thomas Spengler, Jeff Knight, John Thuburn and David James for all their help, encouragement and guidance over the years.

CONTENTS

CHAPTER 1

Launch

Surf's up at Soup Bowl! The pale sand beach is dotted with brilliant white pebbles and strung with dark lines of weed thrown up high by the sea. It is deserted, save for some blackbirds flitting along the waterfront in search of food. Weatherbeaten palm trees flank the beach, clinging to the edge of the land. Giant boulders of twisted and battered coral lie scattered along the shoreline, white foam swishing and swirling around them.

The sea is a deep, dark blue, out beyond the surf at least. Closer in to shore, the colour shifts to vivid turquoise, streaked with the white of breaking and broken waves. This is where the crowds are, such that they are: a cluster of small black dots as seen from the shore, each courting the swell, trying to catch themselves the perfect ride in.

The surfers can bide their time if they wish; the water temperature here never really drops below 25°C. Up above, the sky is often as blue as the ocean, sporting only a few harmless white fluffy clouds. This beach can boast a healthy three thousand hours of sun per year.

Green hillsides climb gently up behind the beach, a mix of grassy slopes and windswept woods. A strung-out village faces the waves, offering tempting makeshift rum bars and fresh fried fish stalls, and also proud white-painted wooden houses complete with decks and the obligatory hammocks.

These are all clearly attractive qualities to those planning their next surf holiday. However, the one key attribute which places it in the top league of exotic surf destinations is that the surf is pretty much always up at Soup Bowl. Year round, only one day in every four is deemed unrideable, a fraction which drops to only one in twenty during winter. While the waves can occasionally be very large, this just provides an added bonus. It is the consistency and reliability of the surf which really distinguishes Soup Bowl.

This blessing is bestowed on the beach by its location. Soup Bowl lies on the uncommercialized east coast of the Caribbean island of Barbados. What it lacks in golf courses and five-star hotels it makes up for in wild and rugged scenery. And wind. The most windward of all the 'Windward Isles', Barbados sits directly in the path of the Atlantic trade winds. Look out east from the beach at Soup Bowl and the next land is the African coast. The trade winds have almost three thousand miles over which to build their strength, and to nurture the surfers' swell.

The consistency of the surf is explained by the consistency of the wind. At Soup Bowl, there will typically be one or two days per month when the wind is from the north, south or west, or maybe even absent altogether. On all other days, the wind will be steady and 'easterly' as we say, i.e. from the east. (In this book we will follow the old convention that names winds by where they come from rather than where they are going, so northerlies for example blow from the north towards the south.) It was this unswerving steadiness of the trade winds that provoked the earliest attempts to apply the concepts of physics to understanding the weather, several centuries ago.

Ten miles from Soup Bowl, on the south side of the island, lies Grantley Adams International Airport. The palm-fronted portal to the island, Grantley Adams plays an important part in our story. It hosts one of about ten thousand weather stations which make up the World Meteorological Organisation's Global Observing System.[1] Each of these stations takes standard weather measurements at least every three hours. Quantities such as temperature, pressure, wind and humidity are recorded and immediately sent to forecasting centres around the globe. Over the oceans, a similar task is performed by a fleet of four thousand volunteer ships and over a thousand automated buoys, robotic drifters which record data while travelling wherever the currents take them.

Forecasters hence receive regular updates during the day and night on changes in surface weather, beamed to them from stations like Grantley Adams around the world. To turn this information into an actual forecast of the future, they need to run detailed computer simulations. The computers literally step forward in time, predicting how the weather changes hour by hour. To make the best forecasts, you need to have not only the best computers, but also the best possible picture of the atmosphere at the initial time. The global network of surface stations provides the foundations for this, but is lacking in one crucial dimension:

the vertical. Information on the state of the atmosphere above the surface is critical for making accurate forecasts, and this is provided by two very different systems.

The first experimental weather satellite was launched by NASA in 1959, but it was not until the late 1970s, a decade after man's first steps on the Moon, that the promise of a near global, space-borne weather observing system began to be fulfilled. A fleet of satellites encircles the Earth today, measuring infrared and microwave radiation as well as visible light. After processing, this data can give estimates of phenomena from the near-surface wind over the ocean up to the elevation of the highest cloud tops. The flood of new and unfamiliar satellite data ultimately led to a step change in our ability to make skilful predictions of future weather.[2]

Satellites, however, can only take us so far. At some point we need actual measurements taken directly from the air at high altitude, in near real time, all around the world. Hence an elaborate ritual takes place twice a day at Grantley Adams, as it does at over a thousand other strategic 'upper air' observing sites around the world: a device known as a *radiosonde* is launched. A radiosonde is a small box of electronics, about as large as a 1 kg bag of sugar, but thankfully only a quarter of the weight. This has to be unpacked, checked and carefully prepared before being attached by a 30 m long string to a helium or hydrogen balloon. In near synchroneity with its colleagues around the world this precious, but ultimately expendable, little box is then released into the sky (see Fig. 1.1).

Over the next one to two hours, it rises up through the atmosphere, measuring temperature, pressure and moisture at regular intervals and radioing these back to the station. Wind data can be obtained by tracking the position of the radiosonde, which can drift over a hundred kilometres from its launch site. Radiosondes typically reach altitudes of around 30 km before the balloon finally bursts, by which time it will have expanded from a 1 m to a 6 m diameter. As it falls, a small parachute opens to make the descent a little safer for anyone below. The vast majority of radiosondes are never seen again, with many ending up in the ocean for example, although if you do find one it will often have a pre-paid envelope attached for its return. A few thousand of these ultimately drop from the sky every single day, yet there has apparently only ever been one fatality associated with a falling radiosonde, and that was during its attempted retrieval from a power line.

Fig. 1.1. An ascending radiosonde, shortly after launch.

Each radiosonde returns vital information about the vertical structure of the atmosphere, such as how the winds and temperature change as you rise up from the surface. Satellite technology has advanced to the point where some information on this can be obtained from space, but the radiosonde data generally gives a more detailed picture, in that information is returned at more closely spaced intervals. A detailed picture of this column of atmosphere can tell us a lot about how it might evolve in the near future. For example, it can tell us how stable the column is: if we were to take the air near the surface and push it upwards, would it just settle back down again, or would it accelerate upwards? Rapidly ascending air masses in the latter case are often linked to thunderstorms and rain; these are examples of *convection*, in which the atmosphere bubbles upwards like the water in a heated saucepan.

Radiosondes have been central to our weather observing system for decades. The design of sensors has been improved over the years, and the introduction of satellite tracking made it much easier to get accurate wind measurements, but the basic design is very similar to that first used in the 1930s. Here at Grantley Adams, observers have been launching balloons since 1965. While the original need for these was for weather forecasting, radiosonde data along with other observations also play a vital role in monitoring our climate. Scientific research on climate change is under-pinned by an ever expanding archive of historical data. Painstakingly compiled datasets of historical observations are used to track the changes unfolding around the globe.[3] In addition, when scientists develop new methods and models for forecasting the weather better, or further in advance, these can be tested to see if they would have been able to 'predict' decades of past weather events.

From late summer through the autumn, the measurements from Grantley Adams are particularly important. This is hurricane season in the tropical Atlantic. These powerful, mature cyclones are readily identified from space by their distinctive cloud-free eye. They are the meteorological equivalents of a shark's fin poking out of the water: an unmistakeable sign of danger. However, even several days before tropical storms reach this stage, ominous cloud structures are identified in satellite images and tracked.

The particular role of the balloons from Grantley Adams is to provide a detailed picture of the storm's environment, i.e. the region of the atmosphere surrounding it. The growth of tropical cyclones such as hurricanes is fuelled by the warmth of the underlying ocean and the amount of moisture in the atmosphere. Other factors can act to destroy a cyclone; in particular, if the wind direction and/or speed changes strongly with height (an effect known as *vertical wind shear*). For one of our weather balloons, this would mean it could get blown first in one direction and then another, as it rises up through the atmosphere. For a tropical cyclone, its base could be blown one way and its top another way. Although seemingly all-powerful, the storm itself can get pulled apart by the surrounding winds. Hence, detailed information on how the state of the atmosphere changes as we rise up from the surface is invaluable.[4]

Tropical cyclones track west across the Atlantic (Fig. 1.2), blown by the trade winds, but then veer to the north as they approach the Americas (under the influence of wind patterns we will learn about in Chapter 3). The regular devastating consequences further north in the Caribbean, and over the United States, are sadly familiar. For the inhabitants of

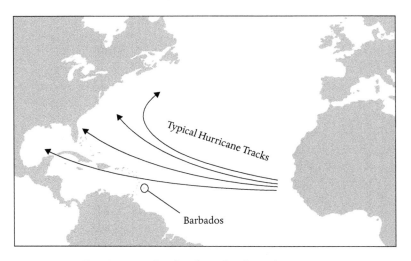

Fig. 1.2. Typical paths of North Atlantic hurricanes.

Barbados, luckily, the most that they normally get is a few very big surf days at Soup Bowl.

As the seasons shift from autumn into winter, relatively little changes in Barbados, although the influx of sun-starved tourists starts to increase. The temperature only drops by a degree or two, and the wind remains steady and from the east. The chances of today being a good surf day rise even higher. For the balloons launched from Grantley Adams, however, change is afoot.

After release, they pretty much always start drifting west, as they do year round, driven by the trade winds from the east. This westward drift slows as the balloons rise, since the trade winds are strongest in the lowest couple of kilometres of atmosphere. But then, something different happens to the winter balloons. From about 5 km up, they turn and then start to go east again, back the way they have come from. In fact, they soon go back past Grantley Adams, and then out over the Atlantic. By the time they reach 10 km, near the preferred flight level for aircraft, they are often moving east at over twice the speed of the trade winds below them.[5] This unexpected change is a case of strong vertical wind shear; it is partly because of this that the hurricane season is now safely over. But this isn't just a local wind feature. It is the beginning of something big; something that for now we will simply call the *jet stream*.

Various analogies have been used to describe the jet stream, each with their own imagery. It is a 'ribbon', 'band' or 'narrow zone' of strong winds, or a 'river' of fast moving air. These are all useful images. Wind is indeed just the movement of the air; this was first recognized by Hippocrates around 400 BC although, as we will soon see, it took a long time for this idea to catch on. The jet stream winds, then, are a strong current of air moving around the globe.

Another useful analogy involves a swimming pool and a powerful garden hose. Put the end of the hose in the pool and turn it on. The water in the pool will be moving around in various ways; perhaps there are swimmers, or some splashing children causing disturbances. From the hose, however, there comes a jet: a fast current of water going in one direction.

In many ways the jet of water in the pool is very similar to the jet stream growing in the sky over Barbados; both are represented in physics by the equations of fluid dynamics. Strange as it may seem, air can be treated as a fluid, just like water. Just as different currents of water swirl and slip past each other in the pool, currents of air dance and twirl around the world in the atmosphere.

When we say the jet is narrow, what do we mean? As it starts to flow across the Atlantic, the strongest winds are typically found in a band a few hundred kilometres wide. This is certainly narrow compared to the size of the Earth, but on a human scale it is very wide: for a pilot aiming for a useful tailwind, it's not an easy thing to miss. We have chosen Barbados as a launch site, but balloons launched anywhere in the Caribbean would have likely had the same fate.

As the balloons rise even further, the wind starts to weaken a little, having been strongest somewhere around 10–12 km above the surface. If the balloons make it up to 20 km the winds will be much weaker. Once past the strongest winds, they have in many ways entered a different world. This is a layer of the atmosphere known as the *stratosphere*. It is eerily quiet compared to the region the balloon has risen through; the ever-changing weather layer known as the *troposphere*. Up in the stratosphere there are no storms, no weather systems, not really any clouds (except some occasionally giving beautiful displays over the poles).

At these heights, the thermometers onboard the radiosondes start to show an important change. The air temperature has been steadily dropping during the ascent, often getting down to below −70°C, but now it begins to rise again. The increase in temperature as you go up is the defining characteristic of the stratosphere. It occurs because of the abundance of ozone in this layer, which strongly absorbs ultra-violet radiation from the Sun and hence warms the air. Since warmer air is more buoyant than cooler air, this means that the stratosphere is very stable. If we took some air here and tried to push it up higher, it would settle back down very quickly. This is what gives the stratosphere its uneasy calm.

Soon after they enter the stratosphere, the balloons from Grantley Adams burst. The parachute opens and the radiosondes begin their slow descent. That's the end for them. In winter, they will end up splashing down somewhere in the Atlantic Ocean.

In this book, however, let's pretend that something different happens. Scientists might charitably call this a 'thought experiment'. It's really just a flight of fancy; one that we're going to take a Grantley Adams balloon on. For the sake of convenience, let's call this balloon 'Grantley'. Hence, he is named after the airport that he was launched from and also, by proxy, a local hero of social justice in the Caribbean.

Let us suppose that Grantley is launched from Barbados in mid-winter but develops a problem. He rises up west through the trade winds, slows, then turns back east. As he rises further, the winds take him eastward faster and faster. Just when he reaches the strongest winds, however, he stops rising. Maybe he loses some hydrogen, or maybe he was too heavy to start with. For whatever imaginary reason, Grantley is now floating perfectly in the sky: with no net force acting on him he neither wants to rise further nor fall down to the ocean. He is trapped in a layer of atmosphere just below the stratosphere. But more importantly, he is trapped in the jet stream, and rapidly heading east.

What will happen to Grantley; where will he go? What climates will he traverse and what weather events might he witness? How does physics govern his path through the skies? What do we know about these bits of the atmosphere, and how? Are they changing, and if so, why? These are the questions we will tackle in this book.

Prior to the early twentieth century, there were no weather balloons and no high altitude flights, so humanity had no direct experience of the jet stream. The trade winds, however, have been a source of wonder and intrigue to scientists ever since their discovery in the fifteenth century. The physics of the jet, it turns out, is intimately linked with that of the trade winds, so this is where the story continues.

Joseph didn't surf. He'd never enjoyed the football or any of the other wild games his contemporaries loved. The hustling and the shouting unnerved him and he felt the pressure keenly. He was never any good at those games, and instead had taken to long, slow walks around the city. As his feet wandered the arcades and piazzas, his mind wandered around his newfound joy. He'd always been a strong student and had worked hard for his tutors. But now he was working for himself, and he was obsessed.

CHAPTER 2

Trades

There is a great quantity of fire and heat in the earth, and the sun not only draws up the moisture that lies on the surface of it, but warms and dries the earth itself. Consequently, since there are two kinds of evaporation, as we have said, one like vapour, the other like smoke, both of them are necessarily generated. That in which moisture predominates is the source of rain, as we explained before, while the dry evaporation is the source and substance of all winds.[6]

Aristotle made countless contributions in many areas, from philosophy and physics to biology and earth sciences. He is widely credited with initiating whole subjects, the field of logic in particular, and for helping to lay the foundations of the modern scientific method. On wind, however, he was quite wrong. He rejected Hippocrates' (ultimately correct) view that 'wind is simply a moving current of what we call air'. Instead, to Aristotle, wind was a dry evaporation, or exhalation from the Earth; an invisible smoke-like substance moving about the Earth.

A triumph of his book 'Meteorology',[7] written around 350 BC, was one of the earliest comprehensive descriptions of the water cycle. This was central to Aristotle's meteorology, with the 'dry exhalation' of wind taking only secondary importance. Wind was merely a by-product of the motion of water around the Earth. Living as he did in the Mediterranean, winds to him came from all directions and were often associated with rain-bearing storms.[8] His ideas would not transfer easily to the tropics which, to be fair, were far outside his own realm of experience.

In ancient China there was also an understanding of the water cycle, but there people were used to a monsoonal climate where onshore winds in summer herald the arrival of the rains. The *Chi Ni Tzu*, published at roughly the same time as Aristotle's Meteorology, was hence much closer to the truth in stating 'wind blows according to the seasons and rain falls in response to wind'.[9] Back in Europe, however, and with Hippocrates' theory long forgotten, it was Aristotle's Meteorology to which fifteenth and sixteenth century sailors turned in order to understand the newly discovered trade winds.

The most celebrated of these, Christopher Columbus, sailed from the Canary Islands on 6 September 1492, in a fleet of three modest-sized ships, determined to find a new westward route to Asia. After an initial two-day period of calm, they made remarkable progress westward across the Atlantic. Columbus' logs for the rest of September become monotonous, recording day after day of calm seas, fine weather and many miles advanced. For example, on 18 September: 'This day and night they made over 55 leagues [190 miles] . . .In all these days the sea was very smooth, like the river at Seville.' A rare headwind on 22 September drew Columbus to comment 'this headwind was very necessary to me, for my crew had grown much alarmed, dreading that they never should meet in these seas with a fair wind to return to Spain.'[10]

By the time land was first sighted in the early morning of 12 October, they had covered a distance of at least four thousand miles from the Canaries to the Bahamas. This was achieved with minimal drama in thirty-six days, averaging just over a hundred miles per day. Luck was clearly on their side; despite crossing the tropical Atlantic in peak hurricane season they had met little more than the occasional shower.

Subsequent voyages would confirm the remarkable consistency of the trade winds (see Fig. 2.1). These were frequently found to blow steadily for many days with not a rain storm in sight. The name 'trade winds' is in fact suggested to originate not from their link to commercial seafaring, but instead from an earlier meaning of 'trade' as similar to 'path' or 'track', suggesting a certain constancy and steadiness.

Aristotle's theory of the wind being driven by the water cycle did not at first sight appear of much use in understanding the trades. However, Aristotle had attempted to explain the horizontal nature of wind, 'for though the exhalation rises vertically, the winds blow round the Earth because the whole body of air surrounding the Earth follows the motion of the heavens'. The trade winds could therefore be understood as the dry exhalation being dragged towards the west by the heavens, as they moved over the stationary Earth.

Columbus himself subscribed to this view. In the letters from his third voyage across the Atlantic, he described how the wind and also the 'waters of the Earth move from east to west with the sky'. In fact, he believed that the whole Caribbean Sea had been formed by the relentless

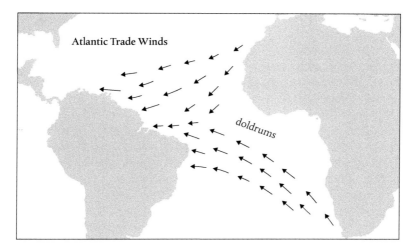

Fig. 2.1. Typical Atlantic trade winds.

onslaught of these westward ocean currents, which had eaten away large tracts of land and hollowed out the sea behind its ring of islands.

It was not until fifty years later that Copernicus published his radical theory, in which the Sun was the centre of the Universe, rather than the Earth. As well as orbiting the Sun, Copernicus claimed that the Earth rotated on its own axis, completing a whole circuit once per day. Despite the huge implications of this theory for the world in general, the sixteenth century picture of the trade winds needed only a minor modification.

It was Galileo who simply turned Aristotle's picture on its head, so that rather than the heavens dragging the wind over the stationary Earth, the trades were a consequence of the atmosphere failing to keep up with the Earth as it spun below. Clearly the surface of the spinning Earth would be moving fastest at the equator, hence it was supposed that it was only there that the Earth moved too fast for the air to keep up. As the Earth spun towards the east, the lagging behind of the atmosphere was felt at the surface as a steady breeze toward the west.

To see the flaw in this theory, let's consider a simple experiment. Imagine we could stop the motion of the Earth, so everything was completely at rest: no spinning planet, no winds, no tides, and so on. Then we give the planet a sudden, sharp spin to set it rotating as usual.

To perform this feat for real would of course be rather hard. Luckily, a similar experiment can be done quite easily with a simulated atmosphere on a standard laptop computer. In the simplest case we don't even need the mountains or continents, just a featureless ball of a planet, the same size as the Earth. Then, what happens if we suddenly start this stripped-down 'Earth' spinning at its usual rate?

Friction will act at the surface of the planet, so that the newly rotating Earth starts to drag the nearest bits of air along with it. The effect of the friction will be gradually spread upwards through the atmosphere, as each bit of air exerts a pull on the next bit of air just above it. As a result, it will gradually accelerate the whole atmosphere, layer by layer. After a few hundred days, depending on the exact model setup, all of the air in the atmosphere will be spinning precisely with the planet.[11] The atmosphere is said to be 'spun up' by this stage, and as it is spinning at exactly the same rate as the planet, no wind would be felt at the surface. If Grantley had been launched into a spun-up atmosphere on such a planet, he would have risen up to his cruising altitude of 10 km and then just stopped there, hanging forever in the sky directly above his launch location.

So, even on this virtual planet, seemingly as smooth as a billiard ball, the gradual effect of surface friction would ensure that the atmosphere would not, in fact, be left behind by the Earth, as it was in Galileo's theory. Hence, the theory is clearly wrong, but it does contain two of the key ingredients that were ultimately used to explain the trade winds. The first of these is the basic fact that the Earth is rotating. The second is the observation that Earth's surface is moving fastest at the equator. To get closer to a basic theory of the trade winds, however, one more key ingredient was needed: the Sun.

By the late seventeenth century the scientific revolution was well underway, and several eminent thinkers were eager to apply the new laws of mechanics to the world around them. The trade winds were seized upon as a particular focus. Due to their steadiness, they were viewed as being simpler in nature than the highly variable winds of middle latitudes, and hence hopefully easier to understand.[12] It was Edmond Halley, the renowned astronomer and mathematician, who first introduced the Sun into the equation.

Early in his career, Halley set his sights on compiling a catalogue of all the stars in the Southern Hemisphere. In 1676, at the age of twenty, he abandoned his degree at Oxford, boarded the *Unity* as a guest of the

Fig. 2.2. Halley's map: the earliest study in global climatology.

East India Company and set sail for Saint Helena in the South Atlantic in order to establish an observatory. Had light pollution been a concern in Halley's day, Saint Helena would still have been a good choice; Angola lies 1200 miles to the east and Brazil 2500 miles to the west. Instead it was the Atlantic clouds which would regularly frustrate him. He returned home two years later having catalogued 341 stars, and with a new-found interest in meteorology, including the trade winds through which he had now sailed twice.

Halley's 1686 paper on the trade winds was built on this experience: 'In this part of the ocean it has been my fortune to pass a full year, in an employment that obliged me to regard more than ordinary the weather'. Much of the paper is made up of a description of the wind patterns of the world, which was as complete as could be realized, including the first attempt ever made to unify these in one map (Fig. 2.2). Arguably this was the first paper in modern climatology, and the earliest ancestor of today's global weather observing system. Halley noted that his map was 'not the work of one, nor of a few, but of a multitude of observers', each playing the role that a satellite or a weather station does today.

In Halley's explanation of the trade winds, Aristotle is laid confidently to rest: 'Wind is most properly defined to be the stream or current of the air, and where such current is perpetual or fixed in its course, 'tis necessary that it proceed from a permanent, unintermitting cause.' Halley was unconvinced, however, by Galileo's theory of the air being left behind by the spinning Earth. One reason he gave for his doubts was the occurrence of 'constant calms in the Atlantic Sea, near the equator' which, according to Galileo, is where the atmosphere should be left behind by the Earth most rapidly. These calms were well known to sailors as the doldrums, which Halley had likely experienced on his journey and were later immortalized by Coleridge in 'The Rime of the Ancient Mariner':

> Day after day, day after day,
> We stuck, nor breath nor motion;
> As idle as a painted ship
> Upon a painted ocean.

Halley's alternative explanation revolved around the Sun. He realized that as the Sun warmed the air, it would expand and become less dense. Then it would rise up, being lighter and more buoyant than the air surrounding it. Nature, of course, abhors a vacuum (Aristotle again) and so other air must move in from the sides to replace it. Couple this to the relative westward motion of the Sun across the tropical sky, and Halley reasoned that most of this air would rush in from the east, i.e. behind the path of the Sun rather than ahead of it. Hence, the easterly trade winds are born.

Halley's name, of course, is well remembered. In addition to a rather special comet, it has also been given to two craters (one on Mars, one on the Moon), an Antarctic research base, a small mountain at the observatory site in Saint Helena, numerous roads and even a pub in southeast London. It has not, however, been associated with the trade winds. That honour has instead gone to a much lesser known Englishman, the London lawyer George Hadley.

'I Think the Causes of the General Trade-Winds have not been fully explained by any of those who have wrote on that Subject'. Hence begins Hadley's groundbreaking paper of 1735. A mere four and a half short pages of simple but elegant reasoning later, and we have the first broadly correct theory for the trades.

As you might have guessed by now, the air does not rush in behind the path of the Sun in the manner suggested by Halley. Instead, it rushes in from the cooler regions to the north and south. The daily migration of the Sun, it turns out, is not so important. What matters is just that there exists a band of latitudes near the equator which are directly under the Sun and hence experience the greatest solar heating. Here lies the air that gets the warmest, and hence this is the air that expands most and so rises. To replace this band of air, cooler air floods in from both north and south.

This, vitally, explains one characteristic of the trade winds that we have neglected to mention so far: they are not generally aligned directly from the east. Instead they are angled toward the equator, approaching it from the northeast and the southeast. This is because as well as being easterly, the wind is also blowing from the north in the Northern Hemisphere, and from the south in the Southern Hemisphere, as the cooler air floods in. These two air masses collide dramatically on, or near, the equator in what is now termed the *Inter-Tropical Convergence Zone*, or *ITCZ*. The ITCZ is the world's greatest rainfall belt, and its seasonal migration sets the pattern of wet and dry seasons across the whole of the tropics.

But why are the trades easterly at all? Hadley's crucial insight here was that as the cooler air flowed towards the equator, it was moving to a latitude where the Earth was moving faster. The Earth spins about its axis: the line running from south to north poles like the core of an apple. What matters here is your distance from the axis. Stand close to one of the poles and you are also close to the axis, hence you complete only a small circle in one day as the Earth rotates. Stand near the equator and, in contrast, you hurtle around a much bigger orbit in the same period of time. In both cases your path traces out a circle, but of a very different radius. It is this radius that is key here.

At the equator the radius from the Earth's axis is a maximum (in fact equal to the radius of the Earth itself). As a result, the air flowing towards it from north or south will have come from a location with a smaller radius, and hence it is moving slower. If the air is moving slower than the eastward moving Earth then, as in Galileo's theory, it is felt at the surface as a wind toward the west, i.e. an easterly wind.

Hadley's theory is essentially correct, although the numbers themselves don't add up exactly. He supposed that the velocity of the air doesn't change as it moves across the latitudes, i.e. that its momentum is conserved. There is, however, a technicality here in that it is not the standard, high school, 'linear' momentum that is conserved in this case, but instead the *angular momentum*. The linear momentum equals mass times velocity. To turn this into the angular momentum we have to multiply it by the radius of the motion (see Fig. 2.3).

The standard example of this is an ice skater pirouetting, or spinning on the spot. When the skater extends their arms out, they spin more slowly. This is because their radius has increased and so, in order to conserve their angular momentum, the velocity must decrease to

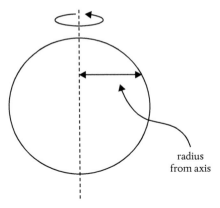

radius
from axis

Fig. 2.3. The radius from Earth's axis of rotation is a key ingredient of the angular momentum.

compensate. For greater applause they have to pull their arms back in again, in order to reduce their radius and hence increase their velocity.

Similarly, as air moves towards the equator its radius from the Earth's axis is increasing, and hence it slows. Hadley neglected this effect, which was unknown at the time. In his theory, the air is moving slower than the Earth when it reaches the equator, just because it has come from a region with smaller radius, but it should have been moving even slower due to the additional effect of angular momentum. This means that the real trade winds are actually stronger than those predicted by Hadley's theory. The difference is considerable: air moving from a latitude of 30° north or south gives an easterly wind at the equator which is twice as strong as in the theory. This, however, is something of a technicality; Hadley's basic idea of how the trade winds work is correct.

What, then, happens to the warm air that rises up from the equator? In this regard as well, Hadley was basically correct. He described how the air 'must rise upwards from the Earth, and as it is a fluid it will then spread itself abroad over the other air, and so its motion in the upper regions must be to the N. and S. from the equator'. This is pretty reasonable. Since the air can't just pile up over the equator, it must spread out again, towards higher latitudes. Eventually it will cool, become denser and less buoyant, and sink back towards the Earth. This is one of the earliest known descriptions of a convection cell, one of the most fundamental of fluid motions.

We now have a picture of a gigantic circulation in the atmosphere. Air is heated most strongly by the Sun in a band near to the equator, so that it expands and rises upward. Needing somewhere to go, this air spreads out northward and southward from the equator, riding over the top of cooler, near-surface air which is flooding in to replace it. Gradually the upper air branch cools and sinks again. Once near the surface, the air joins the queue to head equatorward once more, so that the circuit is complete as shown in Fig. 2.4. What do we call this giant circuit, or convection cell? It is, of course, the *Hadley cell*.

In science it is often the simplest theory to explain something which is most cherished, and the theory of the Hadley cell is no exception. As expressed by Sir Napier Shaw in 1920, 'I have said elsewhere and still hold the opinion that the theory of the trade winds which Halley started and Hadley improved belong to the fairy tales of science because they explain the complexity of nature by a simplicity which is suggestive of a fairy's wand'.[13]

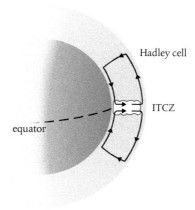

Fig. 2.4. A side view of the tropical Hadley cells (not to scale; the depth of the atmosphere is exaggerated).

Despite this, it took a surprisingly long time for Hadley's theory to take hold. Firstly, Halley's theory had already made it into *Chamber's Cyclopaedia* under the title *'Physical Cause of Winds'*. Secondly, simply because of his name, Hadley was often confused with other people, including Halley and also his own brother John Hadley, who had become well known as the inventor of a popular type of sextant. It was not until one hundred years after his 1735 paper that Hadley's name became widely associated with the theory of tropical winds.[14] Hadley is well remembered today, having given his name to the climate department of the UK Met Office (the Hadley Centre), and to a 120 km wide crater on Mars. Most importantly though, he is remembered in the Hadley cell, a giant system of circulating air in the atmosphere under which 40% of the world's population reside.

Despite his father's elevated position, Joseph's family were not wealthy. All the town gossips knew that they had lost large sums of money on some poor financial speculation. It was clear that Joseph was going to have to work to support them, and his father had been pressing him towards a career as a lawyer. But at college, Joseph found his attention slipping during Latin classes, which until recently had been his favourites. Instead, he was drawn increasingly to the wondrous experiments performed by his new physics tutor. Then, one day, he stumbled on an old scientific paper by Edmond Halley, and he was enthralled. It was not so much the paper's subject that captivated him (optics has deterred many aspiring physics students over the years), more the incredible use of mathematics to explain real physical phenomena.

CHAPTER 3

Spin

It turns out that the Hadley cell is just one example of the powerful effect of Earth's rotation. A striking visual demonstration of this was first provided in 1851 by Jean Bernard Foucault, using nothing more than a simple pendulum. Foucault's pendulum was similar to that which might drive a grandfather clock, except in two regards. Firstly, it featured a carefully designed pivot to allow it to swing without resistance in any direction. Secondly, it was very large: the weight consisted of a 28 kg brass-coated lead ball, which swung at the end of a 67 m long wire mounted high in the dome of the Pantheon in Paris. This combination of length and weight ensured that the pendulum would arc over a very small angle and remain swinging for a very long time with no intervention. It was started as smoothly as possible, in an elaborate ceremony in which a string pulling the pendulum to one side was burnt through, to the delight of the watching crowd.

The swinging pendulum presumably made quite a spectacle, but the crowd's patience had to be tested in order to appreciate the full effect. What was special about Foucault's pendulum was that its direction of motion was changing, drifting ever so slowly clockwise. Foucault had covered the precious marble floor with a wooden platform and, on top of this, a thin layer of sand. A pointer on the end of the pendulum thus traced out the gradual changes in the pendulum's swing in the sand, which must have become apparent to the crowd after around an hour. Let's assume for the sake of argument, that the pendulum started swinging directly east-west. After eight hours it had turned through 90° and was instead swinging north-south. Another twenty-four hours later and the pendulum, still swinging strong, had completed a full circle.

Scientists by this time of course knew that the Earth spun about its axis, but this elegant yet simple experiment was the first clear evidence of this fact that was understandable even to the Parisian public of the nineteenth century. Foucault pendulums are now stalwarts of museums and university science departments around the world, and one has even been constructed at the South Pole. The direction of swing changes faster the closer the pendulum is to the pole, as it feels the Earth's rotation more strongly at higher latitude (for the mathematically inclined, the pendulum turns at a rate of $360°$ per day, multiplied by the sine of the latitude). In agreement with this, the pendulum at the pole rotated one full circle in twenty-four hours, but a pendulum at the equator would not rotate at all. The original pendulum in Paris rotated in thirty-two hours whenever swung, that is until tragedy struck: it was irreparably damaged in 2010, along with the floor of the Musee des Arts et Metiers, when the supporting cable finally snapped.

Convinced, as hopefully you are, that the Earth is rotating, Foucault's experiment tells us something more subtle, something that Hadley had failed to appreciate in 1735. *All* motion on Earth is affected by its rotation, not just that in the north-south direction. In Hadley's theory, the air moving southward towards the equator starts to move westward over the ground, manifesting as an easterly trade wind. If we were to trace its path over the Earth, this air would have turned to the right, drifting towards the west as well as to the south. The pendulum shows us that the same would have happened to air originally moving toward the west: it would be turned to the right, so that it would start to drift north.

The mathematics of this effect had in fact been worked out two decades earlier by another Frenchman, Gaspard-Gustave de Coriolis, in the study of the rotation of water wheels rather than the Earth. The principles are the same, but it still took several decades for Coriolis' discoveries to be fully appreciated for their role in the atmosphere. Today, the deflection of motion by the Earth's rotation is termed the Coriolis effect, and the name of Coriolis is proudly engraved on the side of the Eiffel Tower, along with Foucault and seventy other French scientists, mathematicians and engineers.

A simple demonstration of the Coriolis effect can be performed with a ball and a playground roundabout. (If you're too embarrassed to do it in the park, videos of such antics can be readily found on the Internet.) Sit two people facing each other on either side of the roundabout and set it spinning anti-clockwise. One person then attempts to throw the ball across the roundabout to the other person. Simple as this task sounds, it will actually prove quite challenging, as the ball will appear magically deflected to the right. To reach its target, the ball will have to be aimed slightly to the left.

There is, of course, nothing particularly magical going on here. The ball is moving in a straight line through space, as Newton said it should, and the roundabout is simply turning underneath it. This is the Coriolis effect. It is what we call an *apparent* force, as there is no real force pushing the ball sideways, it just appears so because we are looking at it from the perspective of the spinning roundabout. Similarly, when air moves over the Earth, there is no real force deflecting it sideways, it just looks like that because we are viewing it from the surface of the rotating planet.

By spinning the roundabout anti-clockwise, we have made a situation like the Northern Hemisphere. Looking down on the North Pole from space, the Earth is also turning anti-clockwise. If we instead look down on the South Pole then we see it going clockwise. Spinning the roundabout clockwise therefore resembles the Southern Hemisphere and we see the ball curve in the opposite direction. Hence, a moving object is deflected to the right in the Northern Hemisphere but to the left in the Southern Hemisphere, as in Fig. 3.1. On, or very near, the equator there is no deflection at all.

Thinking back to the surface winds of the tropical Atlantic, as the cool air rushes towards the equator it is deflected by Coriolis. The wind blowing from the north is deflected right, while the wind from the south is deflected left. In both cases the result is a wind blowing from east to west; the mighty trade winds which sustained the voyages of Columbus and countless others after him. The ocean winds are turned towards the Caribbean and away from Africa, leaving a quiet region in the eastern tropical Atlantic where Coleridge's idle painted ships are stranded.

Coriolis himself did not anticipate the fundamental role that his discovery would come to play in understanding the movement of both atmosphere and oceans. This was first worked out entirely independently by the remarkable William Ferrel. At the time that Coriolis' theory was published, Ferrel was a shy farm boy in West Virginia, teaching himself maths from any books he could lay his hands on. When he should have been sorting the wheat for his father he was scratching diagrams on barn doors with the prongs of his pitchfork. Ferrel went on to work as a school teacher, firstly to fund his way through college and then later to support himself while indulging his passion for science.

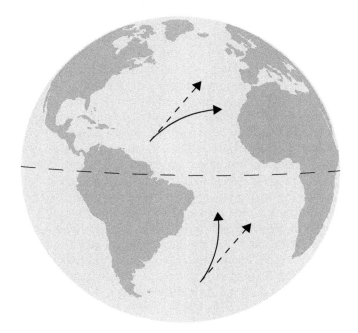

Fig. 3.1. Coriolis turns flow to the right in the Northern Hemisphere and to the left in the Southern Hemisphere.

It was while teaching in Nashville, Tennessee in 1856 that he wrote his 'Essay on the Winds and the Currents of the Ocean'. Ever modest and unassuming, Ferrel published this in a local periodical run by one of his good friends. Hence, this much-celebrated paper is likely the most important work ever to grace the pages of the Nashville Journal of Medicine and Surgery (though perhaps not of such direct interest to its regular readers). In this paper he lays out a theory for what is now known as the *Ferrel cell*: a Hadley cell for the mid-latitudes, in which air rises at high latitudes, on the very edge of the polar regions, before moving towards the equator and then sinking again when it hits the upper branch of the Hadley flow advancing polewards towards it.

While Ferrel did identify this cell for the first time, he unfortunately had an incorrect explanation for it. The physics of this region is so involved that it had to wait another hundred years to be properly explained (or, in our case, until Chapter 8). Instead, Ferrel should be remembered for an elegant, concise and much more mathematical paper just two years later. In four brief pages he derives the maths needed to calculate the deflection experienced by objects moving over the rotating Earth, and applies it to projectiles and pendulums such as Foucault's as well as to air masses flowing into storms, a true milestone in meteorology.[15]

Ferrel essentially realized that what we now call Coriolis is an extra, 'apparent' force, that we need to include in the equations of fluid dynamics that we use to describe the atmosphere (and the oceans). So, what are these equations? Nothing very mysterious in fact. The main equations simply result from applying Newton's second law to a fluid. $F = ma$ is (hopefully) burned onto the brain of every high school physics student, and the basic principle is as simple as that. To apply this to the atmosphere we think of a small bit of air, which we usually term a *parcel*. We add up all the forces that are acting on the parcel, and this sum of forces is the F in our equation. To find out what will happen to this air parcel next, we calculate its acceleration (a) simply by dividing this net force by the mass (m). The acceleration will tell us how the parcel's velocity will change, and so how and where it will move next.

The forces are generally familiar as well. Gravity, of course, is ever present. Friction acts to slow things down, for example as air flows over a rough surface. Hence the wind is usually stronger over the relatively flat ocean than over the land, with its hills, mountains, forests and cities. When you turn on a tap, water is forced along a pipe because the pressure at one end of the pipe is higher than at the other end. Similarly, air in the atmosphere feels a force pushing it from regions of high pressure towards regions of low pressure.[16] To this short list of forces which make up F we simply add the apparent force of Coriolis.

Two more equations are needed as well, to complete the fluid dynamical description of the atmosphere (though further equations are then added if we want to describe factors such as the transport of water vapour and its effects on the atmosphere). We have the first law of thermodynamics, which states that energy has to be conserved: it cannot be created or destroyed, but can be changed from one form into another. This describes how the air is affected by heating, such as by radiation from the Sun or the ground or numerous other sources. Finally, we need an equation to guarantee conservation of air, termed the continuity equation. Air can move around and can change density, but it can never just vanish. It is the continuity equation which dictates the spreading of air away from the equator at upper levels, to make room for more warm air which is impatiently crowding it from below.

The Coriolis effect of planetary rotation is ever present, at least provided you are some distance from the equator, but it is not readily apparent in everyday life. The jet of water from our garden hose is not noticeably deflected to the side, and neither is a ball thrown through the air. Coriolis is not to blame for the difficulties encountered by the England football team in scoring penalties. This was the challenge that Foucault had faced: when the numbers are put into the equations, the Coriolis force is simply much smaller than all the other forces, unless it is accumulated over a very long time, or a very large distance. Foucault captured the effect by designing a pendulum that would swing freely for a very long time. Similarly, moving air in the atmosphere is only noticeably affected by Coriolis if it moves far enough or for long enough. The numbers depend on how far it is away from the equator, but typically it will need to move over a few hundred kilometres or travel for several hours.

A central concept in fluid dynamics, and many other branches of physics, is that of *scale analysis*. By scale here, we simply mean a rough estimate of something, which is hopefully correct to the nearest power of ten. For example, does our radiosonde Grantley weigh 10 grams, 100 grams or 1 kilogram? He actually weighs about 250 grams, so a good figure for the purposes of scale analysis would be 100 grams.

Given the particular scales we are interested in, it is often the case that when we estimate numbers like this, some of the terms in the equations are so small that they can be ignored. As an example, let's think about the increasingly popular sport of kite surfing. Picture yourself on a surfboard out at sea clinging, in all likelihood in desperation, to a large kite flying above you. To try this, you'd better go around the corner from Soup Bowl, to the south coast of Barbados where the waves are weaker.

In a simplistic picture, there are four forces acting on you: gravity, friction, the pull of the kite and the push of the water. The upwards push from the water is normally called buoyancy; it is the same force that kept Columbus' ship afloat and which pushed Grantley up into the jet stream. If all of these forces are in balance then you will not accelerate. For example, if the kite is pulling you west but the friction between the surfboard and the water is as strong, you will keep moving west at the same speed, neither speeding up nor slowing down. If, however, there is an imbalance in the forces, you will accelerate. For example, a sudden gust of wind might lift you briefly up above the waves.

In the opposite extreme, what if the wind dropped so as to barely keep the kite aloft? Then the force of the kite is negligible compared to the other forces. You may as well pack up the kite and go back to riding waves at Soup Bowl, but more importantly, you can ignore that term in the equation. What you've done is make an approximation to the equation, assuming that since the other forces are so much larger, you will only introduce a slight error to your answer if you ignore the force from the kite.

When we apply scale analysis to Newton's second law, we choose the scales of motion that we are interested in, then estimate the size of the different terms in the equation to see if any can be neglected. For example, over distances shorter than a hundred kilometres or times shorter than an hour, we may very well be able to ignore the Coriolis force. For motion much longer than this, however, it will be a very important term in the equation. Essentially, if the Earth has rotated a noticeable amount during the time that something is moving, then Coriolis will be important. So it is crucial for the circulation of air through the atmosphere but is negligible for garden hoses and footballs.

In many situations, in fact, there is a near balance between the Coriolis and the pressure forces. These two will often be much stronger than the other forces and directed against each other, such that they very nearly cancel each other out. The net force will then be very weak, and hence the acceleration will be small. We can then neglect the acceleration as well as the other forces, and in this approximation to the equations, the motion neither speeds up nor slows down. We can use this to explain one of the most obvious puzzles of meteorology.

Weather systems correspond to regions of high or low pressure. If the pressure is high, the weather is likely dry and fine. If the barometer starts falling, the pressure is decreasing and a storm is approaching. This low pressure region is termed a cyclone. A familiar feature of weather maps is that the wind will be blowing around the cyclone, like cars travel around a roundabout. This, on reflection, seems very strange. The pressure force will be pushing the air directly from the region of higher pressure outside the cyclone towards its centre where the pressure is lower. Despite this, the air actually moves at a near right angle to this force, so it doesn't get closer to the low pressure at all.

This situation is often compared to the contour lines linking points of the same height on a map. The wind blows parallel to lines of constant pressure on a weather map (the *isobars*) i.e. circling around 'mountains' of high pressure and 'valleys' of low pressure. To continue the comparison, this is like a river flowing around the side of a mountain, rather than down the slope as it should.

This riddle is solved by realizing that the pressure force and Coriolis are largely in balance, a situation termed *geostrophic balance* (Fig. 3.2). In the Northern Hemisphere, the wind will be blowing anti-clockwise around the cyclone. This is precisely what the word cyclonic means, that something is spinning in the same direction as the Earth. (If you were to look down on Earth from high above the North Pole, a cyclone there would be spinning anticlockwise, and so would the planet itself.) For the air moving around the cyclone, the pressure force will be trying to push it to the left, in towards the centre of the cyclone where the pressure is lower. Coriolis, of course, is at the same time pushing it to the right, and we are in stalemate. The forces balance and there is almost no acceleration.[17]

Similarly, regions of high pressure are termed anticyclones, in that they spin in the opposite direction to the Earth beneath them. Here the wind blows clockwise, the pressure pushes outwards from the anticyclonic centre and Coriolis pushes inwards. All is once again balanced. South of

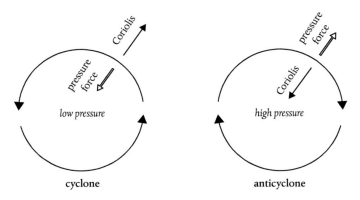

Fig. 3.2. Forces in geostrophic balance for Northern Hemisphere weather systems.

the equator, only the directions are reversed: cyclones spin clockwise for example and Coriolis acts to the left.

The basic mechanics of this had been worked out by Ferrel in his brilliant paper of 1858, though without all the mathematical detail in this case. It had also been discovered independently at about the same time by the Dutch mathematician Buys Ballot, purely from analysis of data from weather stations along the Dutch coast. The result is succinctly written as *Buys-Ballot's Law*: stand with your back to the wind in the Northern Hemisphere and you'll have higher pressure on your right, lower pressure on your left.

In weather systems then, the atmosphere has somehow arranged itself to be largely in balance, with all the forces nearly cancelling each other out. The 'nearly', however, is incredibly important here. Consider what would happen if the cancellation was exact. Coriolis would *exactly* balance the pressure force and there would be no acceleration. In that situation, the velocities of all the individual air parcels would stay the same and nothing would ever change. Today's cyclones and anticyclones would remain constant, still spinning but rooted to the spot forever.

This, of course, does not happen. Storms move, grow and decay, and the weather changes. The evolution of the weather from day to day actually arises from the accumulation over time of all the small differences between these two opposing forces. Hence $F = ma$ becomes a powerful way to predict the future. Suppose that, at a particular instant in time, we know exactly what the atmosphere is doing everywhere. We know the wind speed and direction, the temperature and pressure at every latitude and longitude and every height over the Earth's surface. From this information we can calculate what all the forces are, everywhere, and hence we can calculate the acceleration. This tells us how the atmosphere will change: which winds will strengthen, weaken, or turn; which bits of air will get warmer or colder and which will rise and which will sink. Hence, we can make a prediction of how the atmosphere will be different a short moment later.

This process is the basic method by which weather forecasts are made through computer simulation. As they step forward in time, the simulations are applying Newton's law at each step, calculating all the forces and predicting how the atmosphere will change over the time until the next step (usually less than an hour later). We never, of course, have perfect knowledge of every part of the atmosphere from which the simulations can start shuffling forward, but thanks to the satellites and all the observing stations, we do have some kind of picture. It's a fuzzy picture, with much detail missing, but it's good enough for many forecasts.

Returning again to the tropical Atlantic, the trade winds blowing towards the west are also largely in balance in the same way (a little bit away from the equator at least). As such, they must have high pressure regions to either side, i.e. to the north in the Northern Hemisphere and to the south in the Southern Hemisphere. The trades therefore make up the equatorward flanks of two giant spinning weather systems. Specifically, they are anticyclones, spinning clockwise in the north and anticlockwise in the south. These are the *subtropical highs* which stretch right across the Atlantic Ocean, as shown in Fig. 3.3. They are some of the world's largest weather systems, and only the friction of the relatively rough land over the American and African continents stops them spreading further. In both cases the central point of highest pressure lies about 30° off

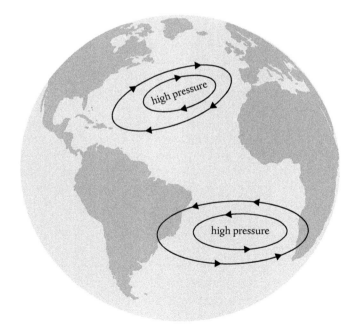

Fig. 3.3. The subtropical highs.

the equator, and it is about these points that the winds circle. North of 30°N then, the wind on average blows from the west instead of the east, although with much less consistency than in the trades.

On 16 January 1493, Columbus set sail from present day Haiti carrying world-changing samples of Caribbean gold and a few kidnapped locals. Steering directly for Spain, his path took him out of the trades and the wind began to blow the ships to the north, and then gradually to the east. Columbus was being blown clockwise around the subtropical high.

All was well until the 13 February, when his logs note an abrupt change. 'From sunset until daylight there was great trouble with the wind, and the high and tempestuous sea.' The weather continued to worsen to the point that, in a desperate attempt to protect his legacy

> even if he perished in the storm, he took a parchment and wrote on it as good an account as he could of all he had discovered, entreating anyone who might pick it up to deliver it to the Sovereigns. He rolled this parchment up in waxed cloth, fastened it very securely, ordered a large wooden barrel to be brought, and put it inside, so that no one else knew what it was. They thought that it was some act of devotion, and so he ordered the barrel to be thrown into the sea.
>
> *The Journal of Christopher Columbus* (during his first voyage 1492–93).

Columbus' luck in avoiding storms had run out. This was no tropical hurricane, however, but a mid-latitude winter cyclone. Larger in size but without such extreme winds, these are very different beasts. We will hear much more of them later, as they are crucial to the story of the jet stream. We leave them for now though, and just note that Columbus of course survived the storm; the ships limped into port in the Azores on the 18 February and the rest, of course, is history.

Scarred by his near miss in the storm, Columbus avoided this route in each of his subsequent voyages, instead choosing to battle directly back east against the trade winds. Over the centuries to come, however, thousands of ships would follow the path of his original voyage clockwise around the subtropical high, in the sinister enterprise known as triangular trade. Slave ships crammed with captives made rapid progress from the African coast to the Caribbean with the reliable trade winds behind them. Laden instead with sugar, the ships headed north and then eastward to Europe, as Columbus had done. On the final side of the triangle they continued their clockwise course around the high, return-ing to the African Slave Coast bearing valuable European goods which could be traded for more, unfortunate, victims. One of the sorriest chap-ters of human history was written with the aid of the Hadley cell and the Coriolis effect.

As we near the end of our own chapter we must rise back up in the atmosphere, with the trade winds and the subtropical high receding below us. What does all of this mean for Grantley? Our hero has been neglected.

We are now in a position to understand the most basic mechanism of the jet stream. It is in fact formed in exactly the same way as the trade winds, just with the directions reversed. We often talk about the Hadley cell as comprising of upper and lower branches. We have discussed the lower branch in detail; air floods in towards the equator from both north and south, and through the turning effect of Coriolis the trade winds are generated. The northern and southern airflows meet and rise up in the ITCZ, the equatorial band of deep thunderstorms. In the upper branch, air is then pushed away from the ITCZ, in order to satisfy the continuity equation, and this flow of air is similarly turned by Coriolis. On the Northern Hemisphere side, this air is moving north and so a *westerly* wind is generated as it is turned to the right. Hence, we have an upper level wind blowing *from the west to the east*, and so the jet stream is born.

The air that Grantley finds himself swept along in likely rose up from the Earth's surface sometime in the last day or two. It might have been over the tropical ocean, either the Atlantic or Pacific, or the Amazon jun-gle (Fig. 3.4). Warmed by the hot surface below, it would have expanded and become less dense than the air above it. The column of atmosphere would have become unstable and the air would have risen rapidly, quite possibly in a dramatic thunderstorm, into the Hadley cell.

While ignorant of the full effect of Coriolis, Hadley and his fellow early meteorologists had in some way anticipated the jet. In the same way that the equatorward moving lower branch produced easterlies, the poleward moving upper branch should produce westerlies. These were variously dubbed the 'west trades' or 'anti-trades'. Owing to their great height above the surface, such winds were not to be directly measured for centuries, but their existence was thought to be consistent with the frequent observations of high cirrus cloud which had been noticed moving eastward in the subtropics.

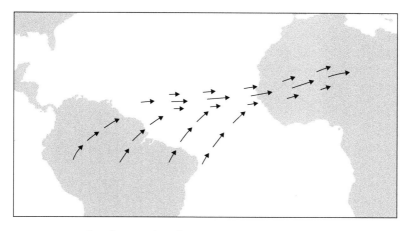

Fig. 3.4. Typical pathways of air flowing north in the Hadley cell and into the subtropical jet.

As the upper air moved poleward, generating the west trades, Hadley realized that it would begin to cool and sink again, spreading the westerly wind down to lower levels. Hence, we have an explanation for the prevailing westerly winds around the north of the subtropical high, the winds which in Hadley's day ushered the sugar-laden slave ships back to Europe. In a satisfying symmetry, his beautifully simple theory could explain both the main winds of the triangular trade.

Except it does not; there are problems that Hadley could never have known about. Sometimes in science it is not as simple as it seems. When the numbers are worked out, the winds predicted by this theory are simply far too strong. Hadley viewed his circulation as extending at least as far north as the region of strongest prevailing westerly winds. However, if we could take some air from over the equator and move it to 45°N, where the westerlies are strongest, it would be shooting by at over 300 m per second. This is over twice as large as the strongest winds ever measured. So what is wrong, should the whole picture be abandoned?

The reality is that the Hadley cell does indeed work much as Hadley imagined, but it just doesn't get that far. The air in the upper branch typically extends no further than 30°N before halting and sinking, and the jet is found here at the Hadley cell edge.

Two questions need answering: firstly, why does the Hadley cell stop here? Secondly, what actually causes the prevailing mid-latitude westerly winds? These winds are essential ingredients of our story, and they certainly aren't 'anti-trades'. For answers to these questions we will have to wait; the prevailing westerlies predominantly live over the oceans, but Grantley is instead heading for Africa.

Joseph was buzzing with excitement as he walked back from posting the letter. For once he felt as if he could run and scream like the footballers. Ever since reading Halley's paper just a couple of dizzying years ago he'd lived for mathematics. Every night he had shut himself away and absorbed all the classic texts at great speed, one after another. Soon, he'd come to desperately desire one thing more than any other: his own name at the top of a paper like Halley's. And now it was done. In his letter he proudly laid out for publication to the world his best work on series. Now, though, he would have to wait for months to hear what others thought of his work.

CHAPTER 4

Contrasts

The pace of activity slows as the early afternoon sun beats down on the marketplace. Stallholders take shelter under the domed concrete roofs and start brewing. The refreshing smell of mint tea fights its way out in tiny bubbles throughout the market. But it will lose the fight; the stench of fish is overwhelming and will stay with any market visitor for days afterwards. Fish of all shapes and sizes lie stacked on the stalls, but the worst smell comes from their freshly removed innards which are piled up on the market floor.

The brief period of calm ends abruptly. Shouts signal the imminent arrival of even fresher fish, and people run for the nearby beach. The white sand burns in the sun, vividly contrasting the brilliant blue of the sea. Colourful patterns adorn the sides of the hundreds of fishing boats lined up along the beach, each one looking like a vibrantly decorated lifeboat.

The excitement is caused by a small gaggle of yet more boats arriving on the shore. They crash onto the beach along with the surf and are set upon by the eager crowds. Each boat disgorges tens of tired fishermen as well as huge quantities of freshly caught fish. The ensuing scene appears chaotic but is actually a well practised and efficient routine. Within minutes the fish are bundled into bags and baskets, or loaded onto waiting donkey carts to be hauled over the rise to the market. A frenetic army sets to work gutting the fish and soon it is laid out for sale and the bargaining begins. What with all the commotion, nobody notices a small white balloon drifting by overhead . . .

Grantley has arrived in Africa, passing over Nouakchott, the capital of Mauritania (and he's far too high to actually be seen, of course). Nouakchott is blessed with white sand beaches and one of the richest fishing areas of the world, but not much else. Perched between the Sahara and the Atlantic, this is not an obvious location for a capital city. A capital was needed though, when the newly independent Mauritania was formed in 1958, and it was established here. The previously nomadic population steadily began to accumulate in the city, arriving in waves during every drought period. With a rapidly growing population and pitifully low annual rainfall, the city's water supply is sustained (for now) by deep wells and a vast underground lake.

Mauritania is critically influenced by the Hadley cell. Directly under the jet stream, it lies in a band of latitudes often termed the subtropical dry zone (or sometimes the horse latitudes). Crucially, this is the region of descent, where the air moving northward away from the equator at upper levels cools and sinks back to Earth. This is dry air and so the sun frequently beats down through cloudless skies. As the air sinks, it also suppresses any infant storms attempting to bubble up from below. This is the most basic mechanism for the formation of the Sahara desert. Most of the world's major deserts are to be found at similar latitudes, in the dry zone underneath the descending branch of the Hadley cell.

At sixty years old, Mauritania remains desperately poor. It faces many challenges, including the ongoing horror of modern-day slavery. This and other illegal activities can thrive in the vast tracts of desert where law enforcement is a daunting endeavour. The same circulation system which drove the triangular trade therefore also provides a wilderness refuge for twenty-first-century slavers.

The name for Nouakchott derives from the Berber 'the place of the winds', but it is not named after the jet which is driving Grantley directly overhead. Down at the surface, life is shaped by different winds. Nouakchott sits at the northeastern edge of the subtropical high which fills the Atlantic basin. Hence the wind is predominantly from the northwest, blowing along the coast and slightly onshore. After leaving Nouakchott, the smell of the fish market will be carried south and ultimately swept into the oceanic trade winds.

During winter, the wind can come more from the northeast; this is the *Harmattan*, bringing very dry and dusty air from the Sahara. In the wetter regions further south, the Harmattan is known as 'the doctor' because of its invigorating dryness. This very dryness, however, has been implicated in fuelling meningitis outbreaks due to its detrimental effect on the lungs. The winds bring a more obvious problem as well: Nouakchott suffers constant attack from a range of giant sand dunes. Comprised of sand driven southward by the winds, the ever-shifting dunes often encroach on the city outskirts. At only a few metres above sea level, the city is hence precariously perched between desert and ocean.

One key benefit is derived from the wind though, and this is Nouakchott's saving grace. It relies on a piece of physics first noticed far from Africa, in as different a climate as Earth can offer. The Norwegian explorer Fridtjof Nansen set sail for the Arctic in 1893, with an audacious plan to reach the pole by freezing his ship into the winter sea ice and simply drifting north with the currents.[18] Nansen was ultimately not to reach the pole, though he did get three degrees of latitude further than anyone else had at the time. Arctic Ocean currents are complex and hard to predict, but he would have stood a better chance of reaching the pole if he had started from Siberia rather than Norway. During his voyage, however, he made a crucial new observation. Nansen noticed that the ice generally drifted not in the direction of the wind but at a small angle to the right of it.

A few years later, Nansen's data made its way to Stockholm and into the hands of a physics student named Fredrik Laurentz Ekman. Ekman would go on to develop a beautiful mathematical theory for how the wind influences ocean currents. This is a cornerstone of today's understanding of both ocean and atmosphere physics, yet Ekman reputedly worked out the equations in just one evening of intense scribbling.

The theory rests on the interplay of not two, but three forces. In addition to Coriolis and the pressure force, Ekman included the effects of friction as the wind blows over the ocean. To see the basic idea, imagine a situation with wind in geostrophic balance (recall that this occurs when the Coriolis and pressure forces exactly balance each other). Then imagine we suddenly add in a frictional force at the surface, as shown in Fig. 4.1. This will slow the wind down and so it will weaken the Coriolis force since this acts on the movement of the air; if the air is moving slower, the Coriolis force is weaker. Suddenly weakened, Coriolis can no longer balance the pressure force, and so the air will feel a net force pushing it towards lower pressure. If we look more closely at weather systems, we see the signature of this effect in the wind, which is angled slightly inward towards the centre of cyclones to make all the forces balance again. Similarly, the wind is angled slightly outwards from anticyclones, as Coriolis is no longer strong enough to balance the pressure pushing outward.

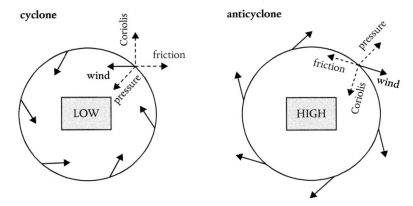

Fig. 4.1. Ekman's force balance for Northern Hemisphere weather systems, with the wind shown as solid arrows and the forces as dashed arrows.

This is all very useful, but Ekman was thinking about the effect of the wind on the ocean, rather than vice versa.[19] We can also apply Ekman's theory to the ocean currents off Nouakchott, which are shaped by the frictional drag of the wind blowing roughly southward along the coast. This sets up a pressure force pushing the water south, and in order to get back into balance the ocean must come up with another force acting to push the water back north. In the resulting balanced state, it is again Coriolis which provides this force. To achieve this, a current must be set up towards the west, so that the Coriolis force, acting to the right of this flow, is pointed north. This is the force that balances the southward push of the wind. But the need for this westward current means that the surface waters of the ocean are drifting offshore, away from Africa. This water has to be replaced somehow, and this happens via constant ascent, or *upwelling* of water from deeper down in the cold depths of the ocean (Fig. 4.2).

The water off Nouakchott is thus surprisingly cool, up to 5°C colder than Soup Bowl even though they are at a very similar latitude. Rising up from the deep, these cold waters are also laden with nutrients, and that brings one vital commodity to Nouakchott: fish. Thanks to Ekman's currents, these are some of the richest fishing waters in the world. They are teeming with life of all shapes and sizes from plankton upwards, and

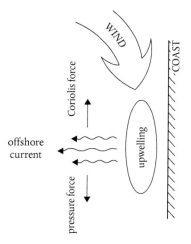

Fig. 4.2. Ekman current in the ocean off Nouakchott.

most importantly fish of all flavours, many of whom are destined to spill their guts on the floor of the city market.

Nouakchott's desert climate contrasts strikingly with that of another capital city further south along the coast: Sierra Leone's Freetown. As the crow flies, only 700 miles lie between the two cities. Yet while Nouakchott is surrounded by desert and receives less than 100 mm of rain annually, Freetown had to be hacked out from the tropical jungle and is doused by over 3000 mm of rain every year. Almost all of this falls during the wet season from May to October, flooding the hillside slums with shocking regularity. This is the region where the arrival of the dry Harmattan wind in November comes as a blessed relief.

Freetown was famously established in 1792 with the arrival of fifteen ships of weather-beaten immigrants from present day Canada. Lured by promises of freedom and land, over a thousand former American slaves crossed all the climate zones that the North Atlantic can offer in order to get here. Upon fleeing the unforgiving winters of Nova Scotia, in which several feet of snow was common, they were battered by severe winter mid-latitude storms, as Columbus and many others had been before them. It was a miracle that no ships were lost, and eventually they limped into the subtropics, grateful for calmer winds and clearer skies. Soon after arrival in Sierra Leone they were met, however, with months of thunder and lightning, violent winds and torrents of rain. As tropical fevers took hold, the snows of Nova Scotia did not seem so bad.[20]

(Grantley, by the way, is eventually headed for Nova Scotia. Or at least, he'll pass just south of it as the jet scrapes the Canadian east coast. Currently heading eastward across Africa, he's clearly taking the long way around, but despite this he'll get there in just under a week. He has, of course, a very good tailwind.)

The striking contrast between the climates of the two African cities arises from their respective positions under the Hadley cell. While Nouakchott is in the descent region, Freetown is firmly under the ascending branch, the inter-tropical convergence zone. Here the whole structure of the atmosphere is determined by the occurrence of deep, moisture-fuelled storms. In contrast to the trade winds, the ascending branch of the Hadley cell is far from steady. Most of the upwards motion of warm surface air is achieved in huge, dramatic thunderstorms, which can frequently take an air parcel right from the ocean surface up into the lowermost reaches of the stratosphere 15 km above.

These storms are particularly frequent over Freetown due to the monsoon effect. As the land warms strongly in the northern summer the air above it rises, sucking moisture-laden oceanic air in behind it, which hits the coast and rises up over the steep hillsides. This combination of warmth, moisture and ascent provides the raw ingredients for storms, which terrorize the slum dwellers today as much as they did the original settlers two hundred years ago.

As important as this contrast between wet and dry climates is for the inhabitants of our two cities, it is a different contrast which is most important for Grantley's journey. As Hadley realized, a crucial factor to consider is the contrast between the tropics and the higher latitudes in the amount of heating received from the Sun. Hadley viewed this in a local way, as he pondered what happens to the most strongly heated air over the equator. There is a more fundamental question, however, which was not pondered until much later: where does all the energy go?

The Earth, of course, is round, and so is warmed by the Sun much more strongly in the tropics than at the poles. In the tropics, the Sun's rays beat down directly overhead, whereas near the poles they hit the surface obliquely, and hence the same amount of energy is spread out over a much larger area. Ultimately, the Earth radiates this energy back to space, so that in the long term the budget is balanced and the planet as a whole neither warms nor cools (providing no-one messes with the budget of course . . .).

The heat loss to space is governed by the physics of so-called *black body* radiation, under which the amount of heat loss depends on the temperature of the surface. While to us the temperature contrast between the tropics and the poles is extreme, in physical terms it is not that large. Temperature in this case should be measured by the Kelvin scale, on which the zero point lies way down at $-273°C$. On this scale, the typical pole to equator temperature contrast is from 260 to 300 K which is a relative difference of only around 15%. The radiative energy loss to space does happen at a slower rate at the poles than over the equator, but not by so much.

The crucial implication of this is that while the planet receives most of its energy in the tropics, it is losing energy almost as much at the poles as it is at the equator. The question then is not really where all the energy goes. That is obvious: energy is constantly streaming from the tropics, where there is a net surplus, towards the poles, where there is a deficit. The real question is how does it get there?

The answer lies in the atmosphere and also, to a lesser extent, in the oceans. This is the most fundamental reason for all the winds and ocean currents on Earth. The atmosphere and oceans have a mission: to relentlessly transport energy polewards, in order to balance the books and to provide energy to the polar regions which can then be lost to space. Most often the mission is accomplished by a simple flow of heat, with warmer masses of air or water moving polewards and colder masses moving equatorward.[21] Without the transport of heat by these circulations, Earth would not even be able to support life as we know it, as the tropics would be unbearably hot and the higher latitudes impossibly cold.

So the temperature contrast between low and high latitudes is of great importance. Let's look closely at how this relates to the jet stream along which Grantley is travelling. The jet, lying at the poleward limit of the Hadley cell, acts as a marker indicating a sharp dividing line in the atmosphere underneath it, which separates the warm tropical air from the colder air to the north. This characteristic is typical of the atmosphere, and indicates the presence of not one, but two complementary types of force balance.

Geostrophic balance, as discussed in Chapter 3, is a balance between the most dominant forces acting horizontally. A similar balance exists in the vertical, and this is known as *hydrostatic* balance. This is simply a balance between gravity and pressure: while gravity acts to pull an air parcel down to the ground, this is opposed by the pressure pushing back up from the higher pressure air below towards the lower pressure above. In hydrostatic balance the forces are actually much more closely balanced than they are in geostrophic balance. When thinking in the vertical, it is nearly always a very good approximation to neglect the other terms in the force balance equation and just keep these two.

We have now identified two dominant force balances: one between horizontal forces and another between vertical forces. We often assume both of these to hold, and then we can merge these two balances neatly into one. This combined relationship shows how the structure of the atmosphere in these two directions is related. Specifically, this gives us a simple equation relating the change in temperature between latitudes to the change in wind with height, which is therefore known as *thermal wind balance*. The directions are as shown in Fig. 4.3: if the temperature is decreasing as you move from the equator towards higher latitudes, then the westerly wind must be increasing with height, i.e. stronger above you than it is below. (As we learnt in Chapter 1, a change in wind with height is known as a vertical wind shear.) We can see that this situation holds

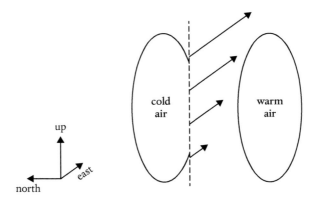

Fig. 4.3. The thermal wind balance relation: if the air gets colder as you go north, the westerly wind gets stronger as you go up.

very well for the jet stream, which indeed gets stronger higher up, as it must do to satisfy the fluid equations.

Grantley is flying along toward the east, in the westerly winds. His path traverses along the edge of a strong temperature contrast in the air below him, with warm tropical air on his right and colder mid-latitude air on his left. By thermal wind balance there must be a vertical wind shear, and indeed the westerly wind gets progressively weaker in each layer down below him. If Grantley were to lose some more hydrogen and fall back to Earth his speed would steadily decrease. For the purposes of our narrative, of course, this doesn't happen.

Another striking example of thermal wind balance is close at hand. Something unusual happens in the lower atmosphere under Grantley's path. Over northern Africa the temperature contrast goes the wrong way, in a local distortion arising from the presence of the mighty Sahara. Particularly in summer, the Sahara is often considerably hotter than the Atlantic Ocean to its south in the Gulf of Guinea. Hence, the temperatures are *increasing* with latitude from south to north, and in thermal wind balance with this the wind must become more *easterly* with height. Near the surface over the southern edge of the Sahara there must therefore exist a reverse, or easterly jet, blowing in exactly the opposite direction to Grantley's course. This jet, which is rather unimaginatively named the African Easterly Jet, has shaped the human history of western Africa for centuries.

The transition zone between the desert climate of Nouakchott and the jungle of Freetown is generally known as the Sahel. It is a giant strip of semi-arid land, comprising a mix of grasses, shrubs and small thorny trees spread out all along the southern flank of the Sahara. Marking the boundary between wet and dry climates, it also lies roughly under the African Easterly Jet, which is pinned by thermal wind balance between the ocean and the desert. This jet is fairly high up, being strongest about three or four kilometres above the surface. For comparison, though, this is only a third of the height that Grantley is cruising at.

As the transition zone between two very different climates, the Sahel is particularly sensitive to variations in climate. Small shifts north and south of the boundary between desert and jungle have had dramatic effects, with periods of drought having the most severe impact. These are not simply occasional dry seasons; droughts in the Sahel are remarkably long. Most recently, for example, the entire period from the late 1960s to the 1980s was unusually dry, ultimately leading to a catastrophic humanitarian crisis.

Historical evidence suggests that prolonged periods of drought have blighted the Sahel for centuries. Civilizations expanded into the savannah regions during the relatively wet periods, and Saharan trade centres such as Timbuktu and Djenné flourished. Eventually, however, an episodic drought period would strike, and as impacts escalated a mass exodus would occur. In previous centuries, droughts would trigger a surge in business for the westbound slave ships, with reports of people even pawning themselves to the slavers in order to save family members. In more recent times, drought has driven thousands of migrants to the rapidly growing slums of Nouakchott and Freetown, amongst many other cities.[22]

To understand why droughts in the Sahel are so prolonged, we first need to think about where the rain comes from. Satellite data in recent decades has shown that very little rain actually comes from isolated thunderstorms, despite their ferocity. Instead, most of the rain comes from larger weather systems, termed *Mesoscale Convective Systems*. These often cover several thousand square kilometres with a deep, thick cloud layer, and can bring steady background rainfall as well as some local downpours due to small storm updrafts nested within the system.

The convective systems themselves don't occur randomly across the Sahel, but are organized by the African Easterly Jet. This jet, like many others, is inherently unstable. We will look at this in more detail in Chapter 7, but the basic concept is that the jet represents a relatively high energy state. One type of energy is related to the speed of the wind, and this is known as *kinetic energy*. With air parcels in the jet moving much faster than their neighbours, there is clearly energy here. Another type is related to the strong temperature contrast across the jet: with warm and cold air masses sat side by side under the jet, there is *potential energy* here too. This is because the cold air is denser than the warm air and so is heavier. It wants to sink down underneath the warmer air, and if it did so the result would be a lower energy state.

There is energy, then, within the jet, and under particular circumstances this energy can be released. Things typically begin with a thunderstorm, a small isolated area of convection. It doesn't give much rain on its own but if it happens to occur very close to the jet, it can lead to something much bigger. The storm is enough to give the jet a slight nudge, and because the jet is fundamentally unstable, that small perturbation can grow. As the disturbance grows, the jet itself gets modified, and starts to release some of its energy.[23] The storm grows and soon the large-scale rains arrive.

Rain over the Sahel is therefore intimately connected to the African Easterly Jet, with storms growing within days or even hours as the jet is perturbed. Why, then, do the drought periods afflict whole decades? The source of those long time scales lies in something much slower than the jet: the ocean. The Easterly Jet itself exists because of the temperature contrast between the ocean and the desert. But the ocean temperature isn't constant, it varies slightly from month to month and also, more importantly, from decade to decade.

Amongst all the oceans, the North Atlantic is particularly noted for its decadal variability, being able to swing to being notably warmer or colder for a few decades at a time. These changes are generally thought to arise from variability in the ocean circulation, the network of slow but steady currents which move warm and cold water masses around the world. The difference is small, with less than a degree of warming or cooling overall, but this is more than enough.

As a result, the temperature contrast between desert and ocean varies from decade to decade, and the laws of physics dictate that the African Easterly Jet must respond to this, by varying its position and strength, in order to maintain thermal wind balance. It seems that two different effects work together; in warm ocean periods, for example, not only is the jet altered but the warmer ocean evaporates more moisture into the air.[24] Over the Sahel there are more storms, and also more moisture that can be rained out. These are the good years.

In the North Atlantic (as elsewhere) the 1960s were a time of great change. Several decades of warm conditions were coming to an end and the ocean was cooling. This was reducing the amount of ocean water evaporated into the air and hence the potential for rain, but more importantly the winds were adjusting to stay in balance with the temperatures. As a result the drought years were approaching.

Other changes were afoot, of course. Swings in Atlantic temperatures are increasingly taking place against a backdrop of global warming, and several studies suggest this is an important factor. While much remains to be understood, there is some evidence that both natural and man-made factors have influenced rainfall over the Sahel in the recent past, and will continue to do so in the future.

The African Easterly Jet is in many ways a smaller, more local version of the hemispheric jet stream which is carrying Grantley rapidly towards Asia. In the story of the Sahel though, the easterly jet which divides the contrasting climates of Africa is of central importance. The vital seasonal rains do not come randomly, but are dictated by dynamical instabilities which grow rapidly, feeding on the energy of the jet. Yet the jet also feels slower influences, adjusting sensitively from decade to decade as the oceans vary, in changes that have shaped the human history of Africa for centuries. Now human activities are in turn influencing the jet, and the future is uncertain. In the easterly jet, in fact, we have the whole story of this book, albeit played out in miniature. The human impacts, however, have been far from small.[25]

Joseph let the paper fall lightly onto the desk. Outwardly quiet and still, his thoughts spun within him. He'd skimmed over much of the text, but the essential point had leapt out, howling at him. It had been done before. His wonderful series derivation, his first attempt at a published piece of mathematics. And not recently, but many years before. His face began to tingle all over, and he could no longer feel the ends of his fingers. He would be branded a copy, the world would think him a cheat. And what, above all else, would Leonhard think of him now?

CHAPTER 5

Waves

When Queen Elizabeth II visited the Swat valley in northernmost Pakistan in 1961, she was so taken by the region that she declared it the 'Switzerland of the East'. With a spectacular combination of lush green meadows, crystal clear lakes and dramatic mountain peaks, it is not hard to see why.

The Swat district is renowned for its productivity as much as its beauty. Wheat, barley, maize and rice all flourish at various elevations, but the area is most famous for the orchards of apple and peach trees and many other fruits that fill the flat valley bottoms. As a bonus the river itself boasts a thriving fish population and supports a healthy trout industry.

Following the Swat river upstream, the orchards give way to narrow, winding valleys, with tidy villages of mud, brick and stone houses nestled in gullies, or clinging to the open hillsides. Some are on such steep ground that the roofs of one level act as a street for the level above. Higher up still are intricately terraced slopes of tiny fields which fade into majestic pine forests, themselves backed by the peerless, snow-capped Himalayas beyond.

Fast forward to July 2010 and the picture could not have been more different. Torrents of water surged down the Swat valley, as northern Pakistan endured its worst floods in over a hundred years. The precious orchards were flattened and over twenty trout farms simply washed away. Even roads and large buildings such as schools and hotels were destroyed when the river finally burst its banks. Also lost were almost all of the fragile, but essential, suspension bridges which hung over the river, connecting settlements on either side. Thunderous landslides struck all over the region, in some cases sweeping whole villages away before them.

Across northern Pakistan, 20 million people were seriously affected by the floods, with over 1700 killed. Two million homes were destroyed or severely damaged, leaving many millions homeless. At its peak, the floodwater covered a fifth of the whole country. Some regions did not fully dry out for six months, and many flood-borne diseases lasted longer still. For the Swat valley, the flood struck a particularly bitter blow. Many here had only recently returned to their homes after fleeing the violence of recent years, which culminated in the Pakistani military re-taking the valley from Taliban control in 2009.

Northern Pakistan is normally relatively dry, receiving only a tenth of the rainfall of the Indian monsoon regions. In 2010, however, the whole of the South Asian monsoon system shifted further inland. The Swat valley is used to storms, even intense ones, so long as they are brief and local. But this time the rain bursts were followed by extended periods of steady, moderate to heavy rainfall, as four separate monsoon surges swept up to the mountains. Half a metre of rain fell in ten days in the Swat valley, which is three times as much as is normal for the whole of July and August combined.

Over a thousand people drowned during this event, but surprisingly, many of these were not in the torrents of the Swat river. They were in a mix of lakes and ponds, rivers and reservoirs in and around Moscow. The Russian population was desperate to cool off, however they could, and escape the worst heat in living memory. Unfortunately the combination of heat, water (and apparently vodka in some cases) proved toxic. Why was Russia sweltering in a heatwave while Pakistan was drenched by flood waves; were these actually part of the same 'event'? To answer these questions we have to follow the jet across southern Asia, and learn about another type of wave.

We rejoin Grantley after his swift ride over the drama of the Sahel below him. He drifts north a little with the jet across Africa, speeding up as he glides over Egypt and the northernmost parts of the Arabian peninsula. By now he's moving at around 150 km per hour, crowded from behind by all the air bubbling upward from the warm tropical Atlantic Ocean. Airliners from the hubs of the Middle East start to shoot past him, keen to share his tailwind. The outbound trip from Dubai to Beijing, for example, will take just over seven hours, while the return leg will be closer to nine.

There's lots of room up there for planes and balloons though, as the jet is still several hundred kilometres wide, so Grantley is in no real danger. Up ahead, however, a rare obstacle is looming. In winter the jet typically passes right over or just to the south of the Himalayas.

At 8848 m, Mount Everest is just tall enough to brush the belly of the jet as it passes. For much of the year the prized summit is blasted with winds typically around 80 miles per hour, giving the mountain its characteristic and photogenic snow plume.

A very special photo of Everest was taken on 28 January 2004 (see Fig. 5.1). From the unique vantage point of the International Space Station, a giant snow plume could be seen stretching east from the peak nearly 20 km towards nearby Makalu. Thanks to a favourable position and a recent heavy snowfall, the astronauts were able to capture the closest thing to a photograph of the jet stream ever taken, as the normally transparent air current was briefly made visible.[26]

Thankfully for would-be summiters, the winds are not always so intense. The jet stream over Asia changes dramatically from season to season, as the Sun shifts about the equator. As the calendar flips through spring and into summer, the Sun moves north in the sky over the Himalayas, and the whole Northern Hemisphere is warmed. The warmest sea surface temperatures are now shifted a little north off the equator and the ITCZ moves with them.

The Hadley cell is now markedly lopsided, with its ascending branch sat north of the equator and very different convection cells to the north and south of this. With the Sun now well north of the equator, the two cells experience different seasons: for the northern cell it is summer while for the southern cell it is winter. The cell on the northern, summer side has relatively little work to do, since the temperature contrasts between

Fig. 5.1. The jet stream carries snow from the peak of Everest towards Makalu.

different latitudes are weaker in the summer hemisphere. This cell is therefore weak, with much less air circulating than before, and it is also shifted a little bit north along with the ITCZ. The southern convective cell, in contrast, is considerably strengthened. This cell is in the winter hemisphere, and so it has a lot of work to do in terms of energy transport as the temperature contrasts are large, with temperatures dropping sharply as we move south away from the equator.

Since the jet is directly driven by the angular momentum transport of the Hadley cell, it shares the same features: the summertime jet is considerably weakened compared to that in winter, and is shifted a little bit towards the summer pole. If Grantley had been a summer balloon he would have drifted past the Himalayas on their northern side, around 500 km north of his actual path, and he would have been going at only half the speed.[27]

As it turns out, May is the window of opportunity for Everest climbers, with most aiming for the second week in May in particular. By this time the jet has weakened considerably and usually shifted north of the mountains. It will stay there and weaken further as summer progresses. But the window is a narrow one, as the surges of the Indian monsoon system are working their way north towards the mountain. By June the snows have arrived and the mountain is again closed.

Even in the May window, however, climbers have to be alert to changes in the weather. There are smaller, local storm systems, which have taken many unfortunates by surprise. In addition, the jet itself does not always play by the rules, as part of its fundamental nature is to be variable. It is not like a river, which on short time scales at least remains fixed within its banks. Look at the jet on any particular day and it will usually be changing somehow, perhaps shifting north or south a bit, or altering its speed or direction a little. Hence, while the jet is usually out of the way in May, climbers have to check it doesn't make an unwelcome return visit. Luckily, the telltale plume of snow from the summit provides a clear early warning of dangerous winds.

If the jet didn't vary, we probably wouldn't care about it so much. It would still impact the surface and mark out the different climate zones of the world, but things would be the same every year. There would be fewer surprises like those which struck Russia and Pakistan in the summer of 2010. It is the variability of the jet which, in combination with other factors, can lead to extreme weather events.

Sometimes the whole jet is shifted a little bit to the north or south. Shifts like this are normally larger over the oceans, where the jet can move by up to ten degrees of latitude. Over Asia though, such shifts will be a degree or two at most. More commonly the jet develops a meander, so that instead of always pointing directly east it snakes its way there, swinging a little north and then south as it goes. This behaviour was very prominent throughout much of July and August 2010: the jet snaked north a little over Turkey, and then swung southward towards Pakistan.

A meandering jet stream is a veritable smoking gun for a very important type of atmospheric disturbance: a Rossby wave (see Fig. 5.2). At its simplest, a Rossby wave is just a sequence of weather systems, one next to the other: cyclone, anticyclone, cyclone and so on. In the summer of 2010, there was an anticyclone over western Russia, then a cyclone over Pakistan, then an anticyclone over China.

You don't actually need a jet stream to develop a Rossby wave, but it certainly helps. When there is a jet, it simply snakes around the weather systems, going with the flow. If Grantley encounters a Rossby wave, he will be drawn clockwise around an anticyclone, which will veer him to the north for a bit. Next he will hit a cyclone and go anticlockwise around it, passing by to the south.

These waves are named in honour of Carl-Gustaf Rossby, the first meteorologist ever to grace the cover of *Time* magazine. The 'likeable, high-spirited and round-faced Swede', as *Time* referred to him in 1956, was the first to derive the simple mathematical theory which is still the basis of our understanding of Rossby waves today. Although he is most

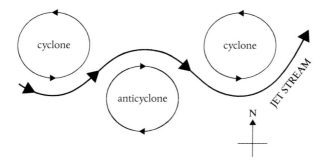

Fig. 5.2. The jet stream snakes around the weather systems in a Rossby wave.

famous for theoretical work such as this, Rossby was a consummate all-rounder, helping to pioneer new methods both to observe the atmosphere and to predict the weather on early computers. He was also an organizational powerhouse, playing a leading role in establishing not one, but three university departments of meteorology (at Massachusetts Institute of Technology (MIT), Chicago and Stockholm) and establishing courses which trained thousands of military personnel in forecasting during the Second World War.[28]

Rossby enthusiastically adopted new technologies, such as the use of balloon-mounted radiosondes to take measurements of the atmosphere. While at MIT he gathered enough balloons to run a month-long test campaign, and ultimately persuaded the US Weather Bureau to begin a national network of regular balloon launches. Rossby's army of prototype Grantleys played a key role in our story: it was these, along with other observations around the world, that revealed to Rossby the abundance of the planetary weather system waves that now bear his name. The early upper air circulation maps which resulted from this effort were also to show for the first time the strikingly global nature of the jet stream itself.

How, then, do Rossby's waves actually work? For any kind of wave we need a restoring force, which pushes something back to where it came from. For the surf waves bearing down on Soup Bowl, this restoring force is just the buoyancy force due to gravity: the crest of a wave which rises up feels the force of gravity pulling it back down again. For a plucked guitar string, it is the tension in the string which tries to flatten it out again.

For Rossby waves, the restoring force comes from something called the *rotational inertia*, which just means the resistance to changes in spin. Pick up a spinning bicycle wheel and you will have to apply some force if you want to make it go faster or slower, or even if you want to tilt it to the side. This is the same phenomenon that we found to be crucial for the trade winds and the jet itself: the conservation of angular momentum. Once applied to a fluid, this will give us the basic mechanism for Rossby waves.

Spin in fluid dynamics is more properly called *vorticity*. A crucial distinction is whether something is really spinning, as opposed to just moving around in a circle. To see this at home, fill a sink with water and then pull out the plug to form a plughole vortex. It will help to give the water a swirl so it has a head start before taking the plug out.[29]

Then, take a matchstick and drop it into the sink a few inches away from the plughole, with the matchstick pointing towards you. If all goes well, the matchstick should be swept around the plughole by the vortex but should keep at the same angle, i.e. pointed towards you (see Fig. 5.3). At this point it is moving in a circular orbit but it is not spinning. (If this doesn't work, try using something bigger like a bath.) Soon, however, the matchstick will be sucked closer in to the vortex and as it does so it will suddenly start spinning around, instead of always pointing in the same direction. This indicates that the water directly over the plughole has considerable vorticity, but in the rest of the sink the vorticity is very small. Away from the plughole, each drop of water is just moving around in a circle, without actually spinning.

In fluid problems, it is often more useful to think about vorticity than angular momentum. Under certain assumptions (there is no stirring or nearby plugholes, for example) the vorticity of a parcel of fluid will not change, so we say the vorticity will be *conserved*. For Rossby waves, these assumptions hold pretty well, and the mechanism underlying the waves is all about the conservation of vorticity.

The final ingredient comes from the shape of the Earth: rather than water moving around in a sink, we now have to think about a layer of fluid clinging to the surface of a spinning sphere. For your next experiment, retrieve your matchstick, dry it off, and take it to the North Pole. Hold the matchstick flat in the palm of your hand and stand completely

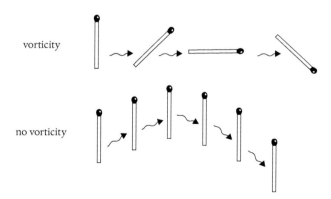

Fig. 5.3. Vorticity, as revealed by a matchstick floating in water. Only if the matchstick itself spins is there vorticity present.

still. Although not immediately apparent, the matchstick is now spinning again, although a bit slower than before. From the perspective of a satellite looking straight down on the North Pole, the matchstick, along with your hand, will be spinning anticlockwise, completing one whole rotation every twenty-four hours. The only reason for this spin, of course, is the rotation of the Earth.

Finally, dust the snow off your matchstick and take it to the equator. Repeat the experiment of holding it flat in the palm of your hand. This time, the matchstick really is not spinning, even from the viewpoint of an overhead satellite. Instead the matchstick, along with your hand and the rest of you, is just moving in a circular orbit through space.

This shows a crucial difference between the latitudes. A parcel of air which is at the North Pole, but at rest with respect to the Earth, still has some vorticity due to the Earth's rotation. A similar air parcel at the equator, in contrast, has none. This type of vorticity is known as *planetary vorticity*. As we move from the equator towards the pole, the planetary vorticity gradually gets stronger (and as it does, the time taken for a Foucault pendulum to complete a full circle decreases).

For an air parcel in the atmosphere, the quantity which is (approximately) conserved according to the fluid dynamics equations is the total vorticity, which is the sum of the planetary vorticity and another component called the *relative vorticity*. This, thankfully, is simpler; it just comes from the motion of the air relative to the surface of the Earth. Large masses of air spinning with respect to the Earth are precisely the cyclones and anticyclones that we have talked about before. The convention in the Northern Hemisphere is that air spinning anticlockwise, i.e. in a cyclone, is said to have positive relative vorticity.[30]

We are now in a position to understand Rossby waves in all their glory. We will begin by taking an air parcel from somewhere in the subtropics and pushing it northwards. By moving the air to a higher latitude we have increased its planetary vorticity, so that it has a bit more vorticity just because of the Earth's rotation. However, its total vorticity can't have changed, as that has to be conserved. The only way for the air parcel to obey this law is if it adjusts its relative vorticity to compensate. In this case the relative vorticity must become negative, so that the parcel spins in the opposite direction to the Earth's rotation. The parcel is then spinning clockwise over the ground, or in other words it has become an anticyclone. Weather systems, then, can be created simply by taking a parcel of air and moving it to a different latitude. Move it towards the pole and it becomes an anticyclone, move it towards the equator and it becomes a cyclone.

To see how this mechanism can form a wave, focus again on the anticyclone which was created by an air mass moving north. This air is spinning clockwise, and will act to drag neighbouring bits of air in

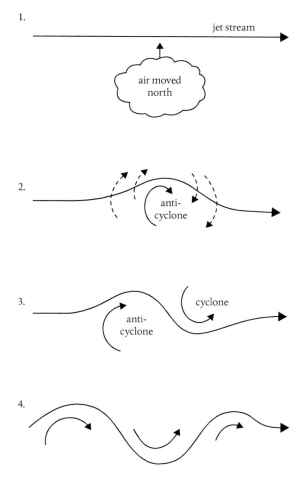

Fig. 5.4. Development of a Rossby wave in the Northern Hemisphere.

the same direction (see Fig. 5.4). For example, the air just east of the anticyclone will be dragged a little bit to the south. But, as we know, air moving south will become a cyclone. Hence the eastern edge of the anticyclone will be eroded away, to be replaced by a cyclone growing on that side. Similarly, the air to the west of the anticyclone is moving north, hence forming a stronger anticyclone there. The overall result is a shift of the original anticyclone to the west, as it is enhanced on its western side and eaten away on its eastern side.

This mechanism explains many things. For a start, it explains why the cyclones and anticyclones tend to form side by side, in what is called a *wave-train*. In addition, as the whole pattern sweeps west, it explains how air parcels are returned to the latitude they came from. We therefore have a wave, complete with its own restoring force. Also wrapped up in this package is movement: Rossby waves by their very nature move to the west.

Or at least, they try to move west. Although the string of cyclones and anticyclones automatically works its way west, it is fighting against the oncoming wind. Away from the tropics, at least, the winds generally come from the west, peaking in the jet stream which is carrying Grantley along. Rossby waves in the atmosphere are therefore always swimming upstream, against the current. Some waves really make it; they move fast enough to beat the jet and actually move westward over the ground. These turn out to be the longest waves, which stretch for thousands of kilometres around the Earth. Shorter waves swim more slowly and so are inevitably swept downstream, so that from the ground we normally see weather systems move from west to east.

In the middle, however, there is a sweet spot. If a wave hits this then it is swimming upstream against the jet at exactly the same rate that the jet is trying to sweep it downstream. There is a stalemate, and with respect to the ground the weather systems just sit there. One region might be deluged under a cyclone, while a neighbouring region perishes under an anticyclone. Hopefully, this scenario is beginning to sound familiar . . .

What happened in the summer of 2010 was the development of a marked Rossby wave pattern across Asia. As in many cases in the mid-latitudes at least, the extreme weather arose because of the persistence of the pattern as the wave remained stationary. For much of July and August the cyclone and the anticyclone hovered over the same regions, so that one grew hotter and hotter while the other got wetter and wetter.

The anticyclone sat over western Russia, with its centre almost exactly above Moscow. Anticyclones are generally associated with settled weather. The flow at the surface is clockwise around the high pressure centre but directed slightly outward, due to the frictional effect that Ekman discovered. This surface air moving outward has to be replaced by air from higher up, and so anticyclones are generally filled with air parcels sinking gradually towards the ground. As in the subtropical dry zone, this descent squashes any convection from the surface and makes the anticyclone dry and often cloud-free. In summer we end up with a warm, dry region of air through which the Sun beats down on the surface.

In 2010 the anticyclone covered a remarkably large area and was stubbornly persistent, and unfortunately arrived at exactly the peak of summer. Maximum temperatures exceeded 40°C in some regions, even reaching 39°C as far north as Moscow. Adjacent nations along the Baltic coast and eastern Europe were also affected, and overall it was estimated that approximately 25% of Europe had not been hotter than this in any of the last 500 years. No wonder people took to the water to cool off.

The thousand deaths from drowning were unfortunately a relatively minor part of the story; some estimates of the total human cost from the heat are as high as 55,000 lives. Over 600 separate wildfires were reported, extending over thousands of square kilometres. The resulting smoke aggravated the usual air quality problem within anticyclones, which occurs as the descending air traps pollutants near the ground. Smog levels in some cities were reported to be as much as eight times higher than usual, and these few weeks alone put a dent of 25% in the total annual crop yield.

Cyclones, in stark contrast to anticyclones, usually contain ascending air and are associated with storms, rainfall and high winds. In 2010 the cyclone lay to the south-east of the Russian anticyclone. It was smaller than the anticyclone but just as persistent, due to the Rossby wave pattern, with dire consequences for the Swat valley and northern Pakistan in general.

These two human disasters were hence tied together in terms of the meteorology by the dynamics of planetary Rossby waves. If there is a silver lining to this cloud, it is that the scale and persistence of these waves offers considerable scope for predictability. Weather forecasts at the time showed high levels of skill even a week ahead of time, successfully predicting various stages of the Russian heatwave and the several distinct pulses of rainfall that struck Pakistan. Steady progress in computer-based forecasting since the pioneering efforts of Rossby and others have made this possible. But obvious and different challenges remain. Despite some forecasts of danger a week ahead, it is unlikely that many in the Swat valley knew what was coming their way.

The jet stream played a very important role in this event, and others like it, because the jet enables Rossby waves to grow more easily. Rossby waves grow because there is a *vorticity gradient*, i.e. a contrast between latitudes where the planetary vorticity is large and where it is small. But the relative vorticity is important too, and in this the jet acts to localize the contrast. It forms a sharp dividing line between relatively cyclonic air to the north and more anticyclonic air to the south, so providing a very strong vorticity gradient. If you could throw a giant matchstick into the jet stream, it would spin in a cyclonic direction if it ends up on the north flank of the jet, and anticyclonically if it ends up to the south.

It would be wrong to infer that the events of 2010 occurred purely because of the jet stream. The jet was important in the formation of the wave, but its meanders were weaker than might be expected given the impacts. If Grantley had been riding the jet across Asia in the summer of 2010 he would have followed a meander north in the anticyclone towards Russia. However, he would only have made it as far as Turkey before swinging back south, with much of the anticyclone and the heatwave still to his north. Although it has been a useful example here, the 2010 event was more than a typical Rossby wave, in several ways.[31]

In particular, the anticyclone over Russia grew to such an extent that it became something we refer to as a *block*. We will discuss this type of event in more detail in Chapter 14, but essentially the anticyclone grew so much that the Rossby wave pattern was locally broken. A whole mass of subtropical air separated off from the wave to swirl around on its own over Russia. The Rossby wave breaking and the formation of the block helped to drag out the lifetime of the flow pattern even more.

Secondly, the cyclonic part of the wave had especially high impacts due to an unusually strong interaction between the different air masses. The rainfall came in several distinct pulses, each of which was triggered by monsoon systems pumping water vapour in from the Arabian Sea and from lowland India and Pakistan. Then each of these rain events was intensified and prolonged due to the cyclonic turning of the air which had moved south from higher latitudes in the Rossby wave.

Finally, was there an extra, added ingredient to the toxic recipe for summer 2010? Record-breaking heatwaves seem surprisingly common these days in many parts of the world; was Russia simply a victim of that trend? Also exceptionally warm that summer was the Indian Ocean, and much of the water that eventually fell on the Swat valley first evaporated into the atmosphere from these warm waters. The story is heating up, but for now we should return to winter and to Grantley, who is bound for Japan.

Far from being discouraged, Joseph redoubled his efforts. He knew he did not want the life of a lawyer, as his family wished for him. He had briefly tasted the excitement and pride of generating his own theorems and could think of nothing else. He was determined to overcome the shame of his first, failed attempt at publishing his own work. He began to work feverishly, late into the night, on a new problem, striding determinedly through a labyrinth of calculus. Finally, he felt he could improve things no further and laid out his ideas as succinctly as possible on paper. With his heart in his mouth he daringly addressed his letter directly to Leonhard, the man he most admired in the whole world.

CHAPTER 6

Storm

'We're not going to listen to you. We're going up there and [we'll] carry out our mission. We'll measure the **** winds and tell you what they are instead of asking you **** what they will be.' Brigadier General Emmett O'Donnell of the US Air Force was clearly unimpressed when the forecasters warned him of potential 170 knot winds over Tokyo.[32] This was the autumn of 1944 and the science of weather forecasting had a long way to go before it would get to the stage where it could predict the Swat valley deluges a week in advance.

This chapter of the Second World War saw high-altitude American B-29s on reconnaissance and bombing missions reach altitudes over 10 km above Japan. This gave rise to the earliest human encounters with the jet stream coming off Asia, often with unfortunate consequences as planes battled against extreme headwinds. One incident occurred on 1 November 1944, as described later by Captain Ralph D. Streakley: 'I found myself over Tokyo with a ground speed of about 70 mph. This was quite a shock, particularly as we were under attack from anti-aircraft guns and were a sitting duck for them. Obviously the head wind was about 175 mph'.

Similar stories also began to emerge in Europe during the latter war years, with planes often finding themselves inexplicably further east than expected, sometimes without enough fuel left to return. Churchill himself received a helpful tailwind on route from Cairo to Tehran for the first conference of the 'Great Three' between himself, Roosevelt and Stalin in 1943.

Encounters with really strong winds over Europe were rare, however, whereas those over Japan were found to be both stronger and more persistent. It was these, therefore, which presented a greater military challenge, and none other than Carl-Gustaf Rossby was brought in to help understand and predict these winds. He is generally credited with introducing the name *jet stream* to describe these, although the equivalent expression *Strahlströmung* had already been used by the German meteorologist Seilkopf in 1939, based on aircraft measurements over Europe in the 1930s. (The term 'jet' was in any case widely used in fluid dynamics before this, such as for a fluid forced out of a nozzle.)

The Second World War stands out in the history of the jet stream as the moment when its existence became not only widely known, but also an urgent challenge for meteorology.[33] It wasn't, however, the first time that the jet had been discovered.

The presence of a jet had in some sense been long anticipated. Hadley and others, for example, had theorized the existence of westerly 'anti-trade' winds at upper levels to balance the surface trades. By the nineteenth century at least, observers had noticed the rapid eastward movement of high-level cirrus clouds and speculated on the strength of wind responsible. It seems likely that others, perhaps much earlier, had gazed up at such clouds and wondered. Some attribute the actual discovery of the jet stream to Wiley Post, the daredevil American pilot who pushed the limits of high-altitude flight in the 1930s. However, not only had the jet been discovered before this, it had been carefully measured and documented.

Mount Tsukuba, 60 km to the north-east of Tokyo, is one of Japan's most visited mountains, as popular for tourists today as it has been for generations of pilgrims. According to local legend, a deity once descended from the heavens and politely asked to stay the night on the majestic Mount Fuji. The mountain considered itself so superior that it was without need of further blessing, but nearby Tsukuba welcomed the deity warmly, even providing food and water. Hence, Fuji was destined to become frigid and desolate while Tsukuba bursts with an abundance of flora and fauna. (An alternative explanation is that Fuji, at 3776 m, is over four times higher.)

The slopes of Tsukuba are mostly wooded, though giant granite boulders lie scattered through the woods. A plethora of tree species blend into each other as you climb the hillside, supporting over seventy species of butterfly and 700 different birds. At sunrise and sunset the mountain often appears a deep purple, due to the dominance of granite and gabbro rock, which contrasts strikingly with the vibrant colours of the surrounding rice fields as they shift from bright green to yellow with the seasons.

But the mountain also boasts its own seasonal displays: bursts of brilliant white and pink in spring as the Japanese cherry and plum trees blossom, and then rich reds and oranges in autumn.

Tsukuba has a distinguished meteorological past stretching back to at least 1902, when Prince Yamashina of Japan founded an observatory at the top of the higher, western peak of the double summit. The mountain had been chosen partly for its location under the route of many passing cyclones.[34] A wide range of meteorological and seismological data were continuously collected and monitored, marking an early step in a concerted Japanese effort to explore the upper atmosphere. The temperature readings from the summit observatory came to be particularly useful for forecasters who were unsure whether to predict rain or snow for the Kanto Plain below.

When Wasaburo Ooishi visited the area a few years later, he concentrated instead on the region south of the mountain. There he found a quiet agricultural landscape of green lawns and sleepy villages nestled between the rice fields. Importantly for him, the area was largely free from obstructions to the airflow, being several miles clear of the mountain and bordered to the east by lowlands that spread right to the Pacific Ocean.

Ooishi's mission was to establish the first true upper air observatory in Japan, motivated in part by the need to develop an understanding of the storms which often struck the region with little warning. Ooishi identified a suitable patch of land which was subsequently purchased by the Japanese Central Meteorological Observatory. Construction of the upper air station was completed in August 1920, with Ooishi established as its first director. Germany had led the field in developing the technology of upper air observation, so Ooishi's project had to wait until after the tumult of the First World War to access the necessary supplies. Observations finally began in April 1921.

To measure the winds far above Tsukuba, Ooishi used pilot balloons, so named as they were often used at airports to determine the wind speed and direction. These were simply large hydrogen balloons which were launched and then tracked from the ground using one or more theodolites; rotating telescopes capable of measuring horizontal and vertical angles. From these measurements, a profile of the wind speed and direction for several kilometres above the surface could be obtained (at least on a clear day; these methods suffer from 'fair weather bias' as you only get weather measurements on days when the weather is good enough).

Over many clear days in Tsukuba, Ooishi carefully measured the winds, often maintaining telescopic sight of the balloons up to heights of around 10 km. Through this painstaking work, he was not only to demonstrate that winds of remarkable strength were to be found in the upper atmosphere, but also to show that such winds were the norm rather than the exception. In winter, the wind tended to strengthen steadily as the balloons rose higher, reaching up to 70 m per second at the top of their range. The wind still strengthened with height in the other seasons, but to a lesser extent than winter. Most of the balloons were of course swept due east from Tsukuba, out towards the great Pacific.

Ooishi was well aware of the significance of his findings, and in 1926 he proudly published his results in a paper summarizing the data from 1288 separate balloon launches. Eager to reach the widest possible international audience, he chose to write his paper in the newly created international language of Esperanto. Unfortunately for Ooishi, this has not yet become popular; by some estimates there are fifty fluent Welsh speakers today for every fluent Esperanto speaker, and Ooishi's paper was sadly ignored for decades.[35]

Within Japan, however, his findings became well enough known to be used for military purposes during the Second World War. It is not clear that this met with approval from the gentle Ooishi, who was as dedicated to tending the gardens around the observatory as he was to launching his balloons. However, over the winter from November 1944 to April 1945, 9000 little warrior Grantleys were launched from Japan on the basis of his observations. These 'Fu-Go' balloons were carefully designed to rise up to 10–12 km and ride the jet stream towards North America, each carrying a small incendiary device.

Ooishi's data have proved highly accurate given the tools available, only very slightly overestimating the speed of the jet at the highest levels. A more serious issue, that the Japanese could never have known at the time, is that the jet stream over Japan happens to be faster than anywhere else along its path. The jet then weakens as it crosses the Pacific; by the time it reaches the other side it has only half the speed it does over Japan. The incendiaries were timed to release after two or three days, by which time the balloons were expected to be over America. In reality, it will take Grantley between three and five days to cross the Pacific, depending on conditions, so most of the incendiary balloons jettisoned their loads harmlessly into the ocean.

Only about 3% of the balloons are thought to have made it to North America, with just two notable strikes. One bomb struck a Sunday School picnic in Oregon, and the resulting six deaths were the only casualties of the war on the US mainland. Another took out the power line to the Hanford nuclear weapons plant in Washington State; the very factory which would produce the plutonium that would devastate Nagasaki.

While the balloon bombs ultimately had a fairly limited impact, the general idea was sound. As he prepares to leave Asia and head out over the vast Pacific, Grantley is not alone. He shares his jet stream with billions of other passengers: a plethora of insect species, bacteria, pollen grains and, increasingly, a heady industrial cocktail of sulphur, nitrogen and carbon-based compounds. Some fraction of this cargo will inevitably manage to ride the jet right across the Pacific, affecting not only Pacific islands such as Hawaii, but ultimately North America as well.[36]

Returning to Tsukuba, the scene today would be almost unrecogniz-able to Ooishi. The city has spread rapidly across the farmland surround-ing the mountain, though careful planning has kept it peppered with trees and grassy parks. In the 1960s, Tsukuba was chosen to host the new national science centre, and since then it has grown into a flourishing city of over 200,000 people. One in every ten inhabitants is a researcher, and the city boasts one of the highest concentrations of research centres and high-tech companies anywhere in the world. Tsukuba is home to particle accelerators and much of Japan's space programme, as well as labs focusing on nanotechnology, agriculture, medical tech and cyber-netics, to name a few. The city now boasts three universities and five sep-arate science museums. Ooishi's observatory is still operational, as the Japan Meteorological Agency (JMA) Aerological Observatory, though the precise site changed in 1975. JMA's Meteorological Research Institute is also based in the Tsukuba Science City, where it has pioneered the development of some of the most detailed computer climate models to date.

Winter is the dry season in Tsukuba. At ground level the prevailing winds are from the north-west as they are over much of Japan. This brings frigid air from Siberia which picks up moisture over the relatively warm Sea of Japan. Heavy snowfalls cover the western side of Japan's mountainous interior as a result, but most of the moisture has fallen out by the time the air reaches Tsukuba on the downwind side of the country. Here the winter days are often cold but clear.

Despite this, snow does fall on the science city with reasonable regularity for a few days every year. The source of this is not the cold, dry air blowing from the mountains. Instead it comes from the sea, as large storm systems scoop up moisture from the Pacific and lift it so high in the atmosphere that it falls out as snow. These are not the tropical cyclones, or *typhoons*, that terrorize the country in summer and autumn. These are weaker but much larger systems termed *mid-latitude cyclones*, and they are exactly the phenomena which Ooishi's observatory was targeting.[37]

Tropical cyclones are bottom-heavy, in that the strongest winds are found near the surface, often with devastating consequences. Mid-latitude cyclones in contrast are top-heavy, with the strongest winds, often over 100 km per hour, up at the level of the jet. A mature mid-latitude cyclone can stretch to around 1000 km in diameter, several times larger than a tropical cyclone. Each storm is a bit like a spinning tower of air roughly the size of Japan, which completes a full revolution in just over twenty-four hours.

Two things are wrong with this picture though. Firstly, due to the aspect ratio of the atmosphere, it's more like a pancake than a tower: 1000 km across but only 10 km high. Secondly, and despite the aspect ratio, there is considerable structure through the depth of the cyclone, with different air masses sliding up and down past each other as they twist around the centre.

At the surface this is clearly noticeable in the pattern of weather fronts that accompany the cyclone, and are familiar features of weather charts around the world. As a cyclone passes, the thermometer will often rise by a few degrees as the *warm front* comes first, in which the warm air from further south is being pulled into the cyclone and starts to slide up over the colder air below. Shortly after this a mass of colder, denser air from the north slides in, slipping under the warm air as the *cold front* passes. Each front involves some air mass or other being pushed upwards, often laden with moisture from the ocean, so this is where the most convection and rain occurs.

Mid-latitude cyclones are common: one passes over Japan on average every three days during winter. Almost as many strike the British Isles, although more often than not the core of the cyclone passes just to the north of Scotland. Britain is still hit by the storm even in these cases, in particular by the fronts which tend to sweep down to the south of the storm centre.

Despite the prevalence of these storms, it was not realized for a long time that these episodes of bad weather extended over such a large, coherent region. Daniel Defoe was one of the first to grasp this, as he compiled 'The Storm'. Defoe's first book was a thorough investigation of the Great Storm of 1703, which decimated forests all over England and wiped out a fifth of the Royal Navy. Over a hundred people died as roofs and chimney stacks collapsed, and an estimated 8000 lost their lives at sea in over a hundred separate shipwrecks. Defoe collected wild reports from all over western Europe: villages torn to pieces, livestock literally blown out of their fields, and windmills spinning so fast that they burst into flame. By piecing these together, Defoe worked out that the storm came from the west and passed over England, France, Germany and then the Baltic, concluding that 'this terrible night shook all Europe'.[38]

Defoe's book has been hailed by some as the first substantial work of modern journalism, but did little to progress the science at the time. Falling back on Aristotle, Defoe speculated that the storm was caused by an exhalation of vapours raised from the Earth by the Sun. He specifically cited the many lakes and inland seas of Florida and Virginia as a likely source, though why he thought these were a better source of moisture than the Atlantic Ocean is not clear. The book, in any case, was a great success. Despite bringing death and misery to many, the storm was something of a gift to Defoe, for whom it provided a suitably apolitical topic to write about after his release from a short spell in prison for sedition.

The first modern picture of an mid-latitude cyclone was drawn at about the same time as Ooishi was establishing his observatory.[39] This was not in Japan, however, but another storm-beaten mountain country: it was produced in Bergen, Norway in 1919 by Jacob Bjerknes (popularly known as Jack, see Fig. 6.1). He was working in a team of scientists under the leadership of his father, Vilhelm, who had been the first to lay out the full set of equations that would ultimately be used for computer-based weather prediction.

In order to support crucial agricultural, fishing and aviation activities during the First World War, the group had reluctantly been diverted from their theoretical work to develop practical methods for storm

Fig. 6.1. Jack Bjerknes at work on the weather charts.

forecasting. Isolated because of its neutrality in the war, Norway was deprived of the weather observations from Britain and Iceland which were used to warn of approaching storms. Instead, the team established their own system of observing stations across Norway, and hence provided vital national weather forecasts. As an added bonus, the detailed data from this densely packed network began to reveal hitherto unseen structures within the storms.

Their detailed study of many storms culminated in Jack's simple picture; the blueprint of an idealized cyclone which was to become a landmark in meteorology (see Fig. 6.2). Near the ground at least, this has all the essential features of a cyclone and nothing else: the anticlockwise flow spiralling inwards, the sharp warm and cold fronts and a sector of warm air to the south, squeezed between the fronts. The term *front* was in fact coined by the Bergen School, in a First World War analogy for the dividing line between warm and cold air masses.

To make such a picture of something 1000 km across that had never been seen in its entirety by any one person took a lot of careful compiling of observations, condensing of previous studies and a piercing insight.

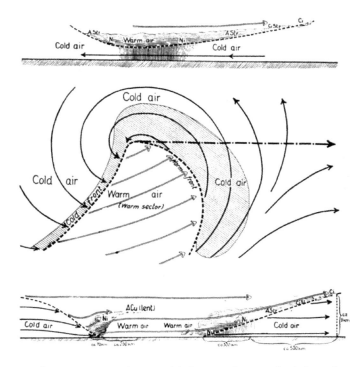

Fig. 6.2. The Norwegian cyclone model, showing the storm both from above and from the side. Both side views are looking from the south, with the top section located just to the north of the storm centre and the bottom one just to the south.

Over forty years later the first weather satellites were to look down on the spiralling cloud patterns and confirm the picture magnificently (Fig. 6.3).[40]

The so-called 'Norwegian cyclone model' was rapidly adopted around the world, as much because of savvy Norwegian publicity as for its elegant fusion of theoretical and practical forecasting considerations. For example, when Vilhelm organized a conference in Bergen in 1920 to share the new cyclone picture, he tactfully split the meeting in half, with the German speakers invited at the end of July and those from other nations in early August. Japan wasn't included but Jack wrote to the Japanese Ambassador, among others, to advertise the Bergen School's results and

Fig. 6.3. A mid-latitude cyclone, picked out from space by the spiralling cloud bands over North America.

also to drum up support for their latest project: the first attempt at a circumglobal weather service.[41]

While the Norwegian model still holds as a remarkably accurate picture, one aspect of their cyclone story has been modified over the years. Bjerknes' team envisioned cyclones as growing from the surface, often forming on an existing front as part of the debris of an earlier storm. They introduced the term *polar front* to describe this low-level temperature contrast on which the storms germinated. Far above their heads, however, lurked another strong temperature contrast that was not to be fully appreciated for decades.[42] It is, of course, the temperature contrast which straddles the winds carrying Grantley around the world.

The jet stream, it turns out, is crucial for mid-latitude cyclones. (And, as we shall see Chapter 7, the cyclones are also crucial for the jet.) Over Africa we learnt how the jet develops along with the strong temperature contrast, where warm tropical air brought from the south by the Hadley cell meets colder air to the north. The technical term we learnt for this relationship between winds and temperature was thermal wind balance. Ten thousand kilometres further on, over Japan, the jet is even stronger.

It has been strengthened by a particularly intense zone of the Hadley cell, itself driven by the warm surface temperatures of the Indian Ocean, the shallow seas of the Maritime Continent, and the West Pacific 'warm pool'. When Grantley passes over Japan he will do so at almost 300 kilometres per hour.

As the jet has become stronger, so has the vertical wind shear underneath it, and hence also the temperature contrast across it, in accordance with thermal wind balance. Grantley, therefore, has never yet been anywhere as unstable as this, with cold dense air below on his left squeezed up close to very warm, light air below on his right. Just as for the easterly jet over Africa, this contrast indicates a great amount of potential energy available, and this energy is about to be released.

Most often the storms start up at jet level, as swirls growing on the jet stream rather than on a surface front. Typically there will be a 'seed': some existing cyclonic eddy such as the debris left by a previous weather system. For Japan, these typically come along the jet from southern China. As the disturbance moves out over the ocean, things start to change dramatically. Suddenly, there is greatly reduced friction compared to that over the rough continent, and something can start to grow at the surface as well.

Growth is particularly strong when there is also a sharp temperature contrast at the surface. In this case, such a contrast is provided by the *Kuroshio*: a strong and narrow current of warm ocean water which passes just to the south of Japan and then heads out east into the Pacific. From the warm to the cool side of this current the temperature drops by several degrees in just a hundred kilometres or so, forming the necessary surface temperature contrast.

Cyclones then develop as the disturbances at the surface and at jet level link up with each other. First, the upper level cyclone reaches down and blows wind across the surface temperature contrast. This in turn generates a cyclonic feature at the surface, in the same way that Rossby waves are driven by north-south airflows.[43] Crucially, the upper and lower cyclones couple together and reinforce each other, often growing rapidly into a full-blown storm in under a day. A key characteristic of the coupling is that the structure has to tilt in the vertical: this pancake of a storm is nevertheless leaning sideways, such that the cyclone centre up at jet level is a couple of thousand kilometres west of that at the surface (see Fig. 6.4).

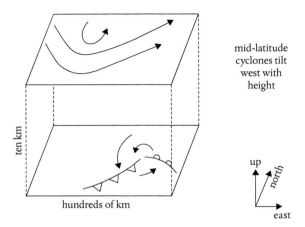

mid-latitude
cyclones tilt
west with
height

ten km

hundreds of km

up

north

east

Fig. 6.4. The alignment between surface and jet-level structures in a mid-latitude cyclone.

This development mechanism would not be worked out until the late 1940s,[44] and so was unknown to Ooishi as it was to the Bergen School meteorologists. However, it shows that the type of upper air observations that Ooishi and others were pioneering were indeed crucially important in providing early warning of developing storms.

Mid-latitude cyclones therefore herald the release of energy that has built up across the jet. As a storm swirls, and warm and cold air masses slide over each other, the cyclone is doing a very important job. It is mixing warm and cold air together, and hence continuing the basic function of the atmosphere to transport energy polewards. Warm air is moved north, cold air is moved south, and these two are eventually mixed. Hence, the temperature contrast and the associated instability is reduced.

By way of analogy, and because Grantley is still over Japan, consider a large pile of rice, freshly harvested from the fields around Tsukuba. Rice from this region was once so prized that for a period before the Second World War it was served in the Imperial Palace in Tokyo. These days, the rice grown here is mostly the relatively new Koshihikari variety; a super-premium short-grain rice which is favoured by many as the ideal grain for making sushi.

Imagine that you are standing beside the rice pile and begin to slowly, but steadily, pour more rice on to the top in a trickle. Initially, the pile is stable, perhaps even resembling the too-perfect cone of nearby Mount Fuji. As more rice lands on the very top, however, it begins to get unstable, as the rice below cannot support the new grains landing above, and the angle of slope gets too steep. What happens next? Well, the whole pile does not collapse, but instead it sporadically adjusts itself every now and then by the action of miniature landslides on its slopes. These move rice downwards, hence reducing the potential energy of the pile since overall the rice is closer to the ground. Energy is released, and the instability is weakened, at least temporarily. The angle of the slope is reduced, back to the point where it is just stable.

Each little landslide of rice is analogous to a mid-latitude cyclone. It does not release all the energy, in the same way as the pile of rice does not suddenly collapse. Instead it moves just enough heat polewards to weaken the instability and make the jet more stable again, albeit only for a few days. Just as the action of many landslides has shaped the pile of rice, the jet stream itself is shaped by the many cyclones that develop over Japan each year.

Before we learn more in the next chapter about how cyclones affect the jet, it would be remiss to leave Japan before mentioning one final cyclone picture. The Great Wave off Kanagawa is one of the most instantly recognizable artworks in the world (Fig. 6.5). Created as a woodblock print by Katsushika Hokusai around 1830, it shows a giant ocean wave towering over three delicate fishing boats, with Mount Fuji lurking in the background. Although often used to represent a Tsunami, the image is actually thought to show an extreme wave generated by the winds of a mid-latitude cyclone. (The clues are in the type of wave, and the fact that it appears not to be tropical cyclone season given the fishermen's clothes and the amount of snow covering Fuji.[45])

Several meanings have been heaped on the wave over the years. A straightforward theme is the battle of ordinary men against great odds, as they paddle audaciously into the wave. Others have read a more subtle message into the picture. The curve of the arching wave is reflected in the curves of the boat and the space beneath it, bringing to mind the opposing, but interlocking forces of Yin and Yang, represented in the good of the fishermen and the evil of the wave.

Fig. 6.5. Great Wave off Kanagawa by Katsushika Hokusai.

To a meteorologist's eye, however, the Great Wave looks a little like a mid-latitude cyclone, as viewed by satellite, with the wave mimicking warm air moving north and wrapping over the cold air moving south.[46] Although Hokusai could never have suspected it, his iconic wave does resemble the very type of storm which created it.

Joseph could hardly look at the envelope which lay waiting on his desk. The merest glance at the handwriting was enough for him to be sure that this was it, a response from Leonhard already. He didn't dare to be hopeful; surely it was not a good omen to hear back so soon. Perhaps Leonhard had not even read his work, and had simply sent a polite reply. He tentatively opened the letter and began to read, but even as he did so he felt the world growing distant around him, as if he were entering the unreality of a dream. The letter practically pulsed with excitement and genuine, honest enthusiasm for his work. His vision blurred with the very beginning of a tear as he reached the letter's final, stunning sentiment: 'I cannot admire you as much as you deserve'.

CHAPTER 7

Tracks

The waves were also high on 13 November, 2002, when the captain of the MV Prestige radioed the Spanish coastguard for urgent help. A mid-sized, single-hulled oil tanker, the Prestige was nearly fully laden and floundering in rough seas off Galicia when one of its twelve cargo tanks was ruptured. Spain, however, refused to offer a sheltered harbour, and the ship was forced back out to sea. Over the following days it was also turned away from both France and Portugal. Six days after the first mayday call, the ship spectacularly split in two and sank, 250 km off the Spanish coast.

Almost half a million barrels of oil were ultimately lost from the ship, in what is widely regarded as Spain's worst ever environmental disaster. Black slime coated the country's northern beaches, and is thought to have killed tens of thousands of sea birds. The thriving local fishing economy was halted entirely until the following summer.

Concerning though this event is, you might wonder why we are discussing it now. From Japan to Spain we have leapt over 9000 miles, fast-forwarding through the whole second half of Grantley's journey. Could the storm which sunk the Prestige really have begun life over Japan? The answer, somewhat confusingly, is yes and no.

Let's return to Tsukuba, which in November 2002 had a cold start to the month. On 8 November, the temperature readings from the upper air observatory jumped suddenly by 10°C but then fell back equally quickly by the next day. These rapid changes indicate the passing of fronts, first the warm and then the cold, that herald a mid-latitude cyclone. Although not extreme at this point, the cyclone was strong enough to bring an early dump of snow. In Gumna Prefecture, just to the west of Tsukuba, the first snow of the season came fifty-two days earlier than normal.[47]

Rewinding briefly to the 7 November, the storm began life as an upper level cyclonic anomaly to the north of Beijing, just a tiny blip on the

weather charts. As is typical, this anomaly moved east across Korea and towards Japan, and started to tap into the instability near the surface over the Sea of Japan. It developed the tell-tale westward tilt with height as the storm grew; by the 9 November, when both fronts had passed and the surface cyclone was out over the Pacific, the upper part of the cyclone was still centred on Japan.

At this point, the storm intensified rapidly. Out over the Pacific the surface friction was low and the strong temperature contrast across the Kuroshio Current provided ample instability. A simple measure of a storm's intensity is the minimum value of pressure in its centre. In this case the pressure fell from 992 millibars (mb) in the early hours of the 9 November to 965 mb a day later. Storms such as this which deepen by at least 24 mb within twenty-four hours are known as *bombs*, for obvious reasons. Luckily this time, the intensification largely happened out over the ocean. After this, the surface cyclone tracked northwards, towards the Russian peninsula of Kamchatka. Up at jet stream level, however, something different was happening. To see what, we have to pause this story briefly to learn more about the link between cyclones and Rossby waves.

Mid-latitude cyclones can sometimes develop on their own, for example, growing off the remnant of some low-level front as envisaged by Jack Bjerknes. This type of behaviour is more common on the eastern side of the ocean basins, over the Gulf of Alaska or the Nordic Seas for example. Near Japan, cyclone growth is normally associated with not just one upper level anomaly, but several, as part of a regional Rossby wave. Cyclones in this area tend to come in wave-trains, each associated with a successive cyclonic part of the Rossby wave. As we learnt in Chapter 5, Rossby waves are effectively swimming upstream against the jet stream. In this case, the waves do not swim fast enough, and instead are blown downstream towards the east.

However, it is not quite as simple as one cyclone in the wave passing by after another. A key feature of atmospheric Rossby waves is that they are *dispersive*. For a simple example of what this means, take a stone and throw it into a reasonably deep pond. After the splash, you should see that the waves spreading out in perfect circles are grouped together in a pack, known as a *wave packet*. Now try to focus your eyes on one particular wave crest in the middle of the pack, and imagine you are surfing this wave across the pond. It should soon be apparent that your wave crest

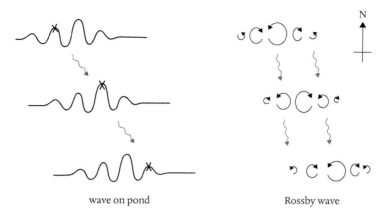

wave on pond   Rossby wave

Fig. 7.1. Wave dispersion: a peak in a water wave moves towards the front of its wave packet, but a cyclone in a Rossby wave moves towards the back of its.

does not stay in the middle of the pack where it started. Instead, it will be making its way forward in the pack, and will also be getting weaker. Eventually, your wave crest will have slipped out the front of the pack and simply disappeared (see Fig. 7.1).

The wave packet represents a little bundle of energy, as bits of water are set in motion and are lifted up and down. The essence of dispersion is that the energy of the wave is moving at a different speed from the individual crests and troughs themselves. In this case, the energy is moving slower than the crests and troughs.[48] Hence, your wave crest has outpaced the flow of energy, so that it has moved out of the front of the wave packet, and has been replaced by new crests that have magically appeared at the tail end of the packet.

The atmospheric Rossby waves that we are interested in behave very similarly, except that the speeds are reversed: this time the energy moves faster than the peaks and troughs. As a result, the individual cyclones and anticyclones that make up the wave get left behind, as the wave packet zips eastward much faster than they do. The weather systems fade away into nothing as the wave packet leaves them behind, but at the same time new ones grow at the front of the packet. Rather than a train of weather systems passing by, a better image is therefore one of a relay race, with each system handing its energy on to the one ahead of it like a baton.[49]

So by the 8 November, when the warm and cold fronts were passing over Tsukuba, the next weather system was already growing: an anti-cyclone out in the Pacific, just to the east. After that came a second cyclone, amplifying rapidly out in mid-ocean on the 9 November. The spacing between the two cyclones was about sixty degrees of longitude. Hence this is often referred to as a 'wave-six' event, because if the wave had spread all around the globe there would be six cyclones and six anticyclones alternating in a row.

The third cyclone was already growing on the 10 November, a further 60° east over North America. Here, it would foster the growth of a multitude of destructive tornadoes. These spectacular whirlwinds are the spitting image of the plughole vortices seen in the kitchen sink. They are much smaller than the mid-latitude cyclones, but are often triggered by these larger low pressure systems. The cyclone over North America spawned no fewer than eighty-three separate tornadoes across seventeen states, in one of the strongest fall-season outbreaks on record. A total of thirty-six people lost their lives, mostly in Ohio and Tennessee. Many of these were in mobile homes, cars, and even whole houses which were simply blown away by the wind.[50]

The fourth cyclone in the pack was developing while the tornadoes still battered the US. This one was again 60° east of the previous, and would rapidly grow into an Atlantic storm powerful enough to sink an oil tanker. The storm was still growing on the 13 November, when the first tank on the Prestige was ruptured, but it would linger and intensify off the Spanish coast for several days while the ship floundered. One final blow remained, however: drifting slowly east, the cyclone eventually hit Europe itself, bringing flooding and mudslides to much of northern Italy and the Swiss Alps.

So it was not, in fact, the same storm that grew over Tsukuba which sunk the Prestige, but it was the same Rossby wave packet. The energy associated with this disturbance took just a week to traverse the jet from Japan all the way to Europe, moving several times faster than each of the individual cyclones which made up the wave-train.

Although much more mobile and dynamic than the wave which caused so much misery in 2010, wave packets like this still bring with them some benefit in terms of potential predictability. Once triggered, any forecasting system worth its salt should have reasonable skill in foreseeing the subsequent development of the wave. Predicting the explosive growth of the initial cyclone and the triggering of the wave is harder, but still within the power of state of the art systems in many cases.

The November 2002 wave packet was one of several events which prompted the establishment of The Observing System Research and Predictability Experiment (THORPEX) international research programme. This endeavour saw researchers from around the world collaborating to advance the fundamental science of observing, understanding and predicting extreme weather events.

While the ultimate consequences of this event were particularly devastating, explosive storm growth near Japan is not unusual. The combination of a strong jet at upper levels and the Kuroshio at lower levels makes this one of the most common regions of storm genesis on the planet. The resulting cyclones drift eastward across the ocean and spawn new storms ahead of them via Rossby wave dispersion. The strip of ocean stretching east from Japan to North America is hence battered by storms, with typically two or three to be found somewhere in this region at any one time. This zone is known as the North Pacific *storm track*.

Each of the major ocean basins has its own storm track. These are key components of the global climate system, taking over from the Hadley cell in transporting energy poleward in order to balance the planetary budget. This task is achieved by the action of storm after storm mixing up warm and cold air masses. The instability driving the storms comes from the sharp temperature contrast across the jet, and this contrast is weakened as the storms do their work.

This is just the same as the behaviour of the rice pile: when the slope gets too steep, the grain landslides act to lower it back down again. Just as the landslides affect the slope of the rice pile, each storm affects the region of atmosphere that it passes through. As Grantley starts to drift out over the Pacific, he therefore encounters a dramatic change; the jet stream will never be the same again.

Recall that the flow is largely in thermal wind balance, so that the temperature contrast in the horizontal is tied to the vertical wind shear (i.e. how sharply the wind changes as you move upward). So far Grantley has traversed along the top of an abrupt boundary between relatively warm air on his right and relatively cold air on his left. In balance with this, there has been a strong vertical wind shear, as the wind at altitudes below him has been much slower than at his level. Now that he is out over the Pacific, however, the storms have weakened the temperature contrast. It is still warm on the tropical side, and cold on the polar side, but the contrast is much weaker. In balance with this, there must be a weaker vertical wind shear, and this is how the storms have changed the jet.

Grantley passed over China at around 60 m per second, but the jet stream winds there were limited to upper levels, with no hint of them at the surface. The jet traversed high above the whole Asian continent, only just clipping the tops of the highest mountains. By the time he reaches the mid-Pacific, however, Grantley is riding on the top of a much deeper structure. His speed has dropped by half by the time he is level with Hawaii, which lies several hundred kilometres to his south. Yet at the same time the wind in the lower few kilometres of atmosphere has strengthened, with westerlies extending all the way from Grantley's level down to the very surface. Hence, the wind changes only gradually below him, level-by-level, and so the vertical wind shear has been reduced.

Right down at the ocean surface the wind blows at around 6 m per second on average, which is weaker than the trade winds, and opposite in direction. But in contrast to the trades, these are highly variable winds; mild westerly on average but this average hides dramatic shifts in wind strength and direction from day to day, as each individual storm swoops past. Mid-latitude westerly winds like these occur over all the major oceans; these are the prevailing westerly winds familiar to many. In the Atlantic, it was these winds that helped Columbus home, albeit with a severe buffeting along the way. Similar winds encircle the globe over the Southern Ocean: the dreaded Roaring Forties, so-named for terrorising ships between 40°S and 50°S.

By weakening the upper wind and strengthening the lower wind, the storms have effectively spread the jet downwards to the surface, as in Fig. 7.2. This deep westerly wind structure is so different from the current that Grantley has followed over Asia that it is clearly a different type of jet. It has acquired characteristics of what is known as an *eddy-driven jet*. 'Eddy' is just the fluid dynamical term for weather systems; cyclones and anticyclones can be viewed as eddies swirling in the fluid, like those we might see in a fast-flowing river.

Crucially, as suggested by the term 'eddy-driven', the weather systems have a net driving effect on the jet. As well as spreading the wind downward, the total momentum in the jet is actually increased because

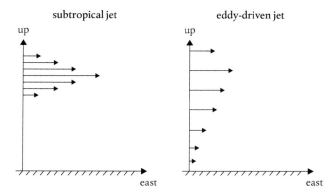

Fig. 7.2. Comparison of a shallow, upper-level subtropical jet with a deep eddy-driven jet.

of the eddies. This happens during the decay phase of cyclones and anticyclones: as the weather systems fade away, much of the momentum associated with their swirling motion is fed into the westerly wind, accelerating the jet.

Over the Pacific then, there are two distinct mechanisms giving rise to the jet. First, there is the usual momentum transport associated with the Hadley cell in Chapter 3, and secondly there is the effect of the eddies pumping momentum into the jet. A jet forced by the Hadley cell is known as a *subtropical jet*. All across Africa and Asia, the jet Grantley has followed has been purely subtropical in nature. Over the Pacific, the eddy forcing kicks in and the jet becomes a hybrid of the two types. Grantley is effectively riding two different jets at this point, they are just merged together.

The eddy-driven surface westerlies have had profound consequences for human civilizations all around the Pacific. For example, they presented an unsurmountable obstacle for sail-powered ships wishing to cross from North America to Asia. This is thought to be one of the reasons why China and other eastern cultures developed so separately from those in the west. The main oceanic trade route to China was from the Indian Ocean, though this had its own challenges, demanding careful timing to catch the favourable monsoon winds. Only hardy Russian fur traders crept around the northern edge of the Pacific storm track instead, shuttling back and forth between Siberia and Alaska.

It was fur trading, though, that eventually opened the Pacific up to trade. Cook first visited Hawaii in 1778, on the third of his great voyages, and then went on to tour the north-east Pacific, collecting furs and other loot along the way. Cook himself met a grisly end while calling again at Hawaii on the return, but the remnants of his expedition made it to China where they managed to sell on their fur for great profit. Soon a regular trade was established, first by British and later American merchants, who stripped Alaska and western Canada of its precious fur and ferried it to China, now by going south to Hawaii and riding with the benign easterly trade winds.

Pacific trade shipping today is radically different. Mammoth container ships laden with manufactured goods from Chinese factories steam east to North America. Driven by the need for efficiency, these ships do not have the luxury of avoiding the storm track. Instead they follow the mid-latitude westerlies, attempting to dodge the storms as best they can. Often, however, the storms are unavoidable.

For example, the container ship 'Ever Laurel', of the Evergreen Marine Corporation, was halfway from Hong Kong to Tacoma, Washington State, when it was struck by a severe cyclone in January 1992. The storm peaked on the 10 January when the ship's log recorded rolls of up to 55 degrees either way; incredibly this means the ship's deck was closer to vertical than horizontal at times. Of the few thousand 40-foot containers on board, twelve broke loose during the storm and crashed into the ocean. Of these, one famously contained a shipment of 28,800 'Friendly Floatees' plastic bath toys, an even mix of red beavers, green frogs, blue turtles and, of course, yellow ducks.

This created a giant, accidental fluid dynamics experiment which attracted worldwide interest in where the toys might be taken on the ocean currents. The first began to wash up on Alaskan beaches after ten months, about 2000 miles from where they were dropped. Many more would be found by a dedicated army of beachcombers over the years, curiously appearing in batches every two to three years. Computer models at the time predicted that some should instead move southward to Hawaii, while others would track even further north, escaping the Pacific through the Bering Strait and the icy channels opening up in the Arctic. Some finds have indeed been suggested as far away as the British Isles, but sadly there seems considerable doubt about this, with no proven examples of the toys appearing outside the North Pacific.[51]

While not the global sensation many hoped for, the Friendly Floatees reveal the profound impact that the mid-latitude westerlies, and also the trade winds, have on the ocean. Underneath these winds lie powerful ocean currents, with colossal masses of water flowing around the basin. Initially, the toys floated high on the water and were largely blown along by the wind. Soon, however, they became waterlogged and partially submerged, but then the ocean currents took over and carried them onward in the same direction.

The toys were caught in a current known as the North Pacific sub-polar gyre, a giant anticlockwise circulation filling the whole width of the northern Pacific and reaching from the mid-latitudes up to Alaska. Stuck in this trans-ocean whirlpool, the toys moved east towards North America but then swung north, where some of them (only 3% in the end) washed up on the beaches of Alaska and the Aleutian Islands. Some were swept twice, or even three or four times around the gyre, which is thought to explain the timing of the finds.[52]

Had the spill occurred a few hundred kilometres further south, the toys would have approached the US, as before, but then been swept the other way, southward around the clockwise subtropical gyre and possibly to Hawaii. The subtropical gyres have the dubious honour of being the dirtiest regions of the ocean. Surface water in these gyres slowly converges into the centre, so that any plastic or other floating debris collects inside the gyre. Although often called the Great Pacific Garbage Patch, most of the garbage here has been broken up into tiny, near-invisible plastic particles by the waves.

The ocean gyres are essentially *wind-driven* features: they owe their existence to the pattern of ocean winds, with westerlies in the mid-latitudes and easterly trades in the tropics. The water underneath is driven roughly in the same direction as the wind, but this seemingly simple relationship actually arises from a complicated dynamical balance. We will delay the discussion of this for a while, until Grantley meets the Atlantic Gulf Stream in Chapter 11, but for now we just note an important feature of the gyres: their asymmetry. Across most of the subtropical Pacific, there is a net southward flow of water. But of course, what goes south must come back north, and it turns out that all of this return flow has to occur on the very western edge of the Pacific. This explains the strong, warm, current of the Kuroshio which flows northward from the tropics to Japan and then out into the ocean interior (see Fig. 7.3).

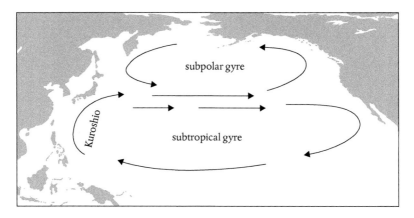

Fig. 7.3. North Pacific ocean currents.

Another complexity in this part of the climate system now becomes apparent. The 'bomb' cyclones developing off Japan are strengthened by feeding off the energy contained in the strong temperature contrast across the Kuroshio. Yet this ocean current is driven by the surface westerly winds which are themselves driven by the storms. The winds and storms would be there to some extent without this feedback from the ocean, but it clearly acts to intensify the situation.[53]

As Grantley speeds across the Pacific, with such complications far below him, one final piece of our story is found here, in a key location just to the south of his path. Mauna Loa is a giant but gently sloping volcanic peak on Hawaii's Big Island. Remote and isolated from major sources of pollution, Mauna Loa was chosen in 1958 to host a vital new atmospheric observatory. This audacious collection of huts and masts, perched on the barren mountainside at over three thousand metres above the Pacific, has taken one of the longest and most consistent measurements of carbon dioxide in the atmosphere.

At its inception in 1958, the observatory recorded a carbon dioxide concentration of 316 ppmv (parts per million by volume). This has increased every single year since then, to over 400 ppmv today. Our best estimate of pre-industrial conditions is 280 ppmv, indicating an increase of over 40% since the dawn of the industrial revolution. Ice core records stretching back over 800,000 years suggest a natural range of variability between 180 and 300 ppmv, so that the man-made contribution to atmospheric carbon dioxide is already as large as the natural variations which took place over many thousands of years.[54] The cause, of course, is the relentless burning of fossil fuels, with global daily consumption of oil alone presently at just under one hundred million barrels. This equates to about two hundred times the cargo of the MV Prestige, burnt up and released into the atmosphere every single day.

With the strong support of Leonhard behind him, life changed quickly for Joseph. Before long he had a job, not in the dreaded law profession, but as a professor of mathematics. He had the time and space to concentrate on his work, and he was ambitious. He was inspired and amazed by stories in which the complexity of the world had been tamed and conquered by a multitude of elegant equations and laws. But he had great plans of his own, determined to impose a new order. The task which Joseph set himself was nothing less than to unite all of these different principles, and hence remove the art often needed in problem solving. Wonderful though each and every proof was, his dream was to reduce everything, as much as possible, down to a few powerful, all-encompassing formulae.

CHAPTER 8

Experiments

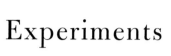

Although it measures only thirty miles in length, this is enough to make Guam the largest of the islands in Micronesia, an archipelago scattered wide across the vastness of the western Pacific. Coral reefs, palm trees and white sand beaches abound, but unfortunately for Dave Fultz, he arrived at the peak of the wet season. From July to October the island lies directly under the ITCZ, bringing deluges on three days out of every four. Monthly rainfall totals regularly exceed 400 mm, roughly four times what Fultz was used to in his home town of Chicago.

It was the summer of 1945 and Guam had been reclaimed from the Japanese for a US air base. This put Tokyo, 1500 miles to the north, just within range of the American B-29s. Fultz had learnt meteorology at Rossby's school in Chicago, and it was Rossby who sent him as a consultant to the Twentieth Air Force Forecasting Section in Guam, to deepen understanding of the newfound high-altitude winds which were hampering operations. During his time on Guam, Fultz helped to prepare forecasts and analyse the incoming data, including looking into reports of the Japanese balloon bomb campaign. He would come to be known as a man of few words, but a compulsive note-taker with an obsession for accuracy.

Fast forward to today and Guam remains a strategic location, with 29% of its land area occupied by US military base. Thankfully, at least some international relations are improved, with Japanese visitors making up 75% of the island's tourist numbers. In today's history books, the name Fultz has a strong presence in the story of the jet stream, although not from his time on Guam. On Fultz's return to Chicago, Rossby issued him with a challenge that steered him into the field that would be his life's work.[55] The new generation of upper air observations was revealing the power and reach of the jet stream for the first time. To explain the existence of the jet would take all the tools available to meteorologists,

87

including simulation. The challenge that Rossby gave to Fultz was to see if he could replicate the jet stream in a laboratory.

A few scientists had attempted to simulate the atmosphere in this way before. The earliest was the German meteorologist Friedrich Vettin, who as early as 1857 had built rotating air tanks in the laboratory. As befitted the time, he tracked the motion of the air in his tank using cigar smoke, revealing slowly overturning billows similar to the Hadley cell. Rossby himself had dabbled in fluid experiments in the 1920s, in some free space he found in the basement of the US Weather Bureau. But now, with the discovery of the jet stream, such experiments had a clear target. Fultz, with his attention to detail and completeness, would be the one to hit this target.[56]

The apparatus that Fultz assembled consisted of two glass hemispheres, one of which was slightly smaller and nested inside the other (see Fig. 8.1). The gap between these could be filled with liquid, which would represent the atmosphere, and then the whole apparatus was rotated around its vertical axis at anything up to sixty revolutions per minute. With a 10 cm radius, Fultz's planet was about as large as a football, and

Fig. 8.1. Fultz's hemispheric shell apparatus in the laboratory.

the atmosphere was a mere 1.6 cm deep (although relatively speaking this is actually incredibly deep, when compared to the aspect ratio of Earth's atmosphere). To measure the movement of the water within this shell, Fultz used tiny pellets with density engineered to float perfectly within the water. A camera was rigged to photograph the experiment precisely once per revolution, and then individual pellets were traced by hand from frame to frame in order to provide quantitative measurements of the flow speed.

In one aspect, the experiment differed strikingly from Earth's atmosphere: heating was needed, to initiate convective motion and circulation, but for ease of construction the heating element was placed at the bottom of the hemispheres. This setup resembles a planet with the Sun sat directly over the South Pole, as opposed to the equator, but it is a testament to the insight of Fultz (and probably Rossby) that this actually did not matter. Invariably, the fluid was found to develop a systematic west-to-east flow in the 'mid-latitudes' of his tank. When scaled back up to Earth-like dimensions, this was even found to be of comparable speed to the recent observations of the real atmosphere. Incredibly, Fultz's model produced a jet stream similar to our own, even though his planet was heated at the pole rather than the equator. Subsequent experiments confirmed that only two ingredients were necessary for the jet to form: rotation and mixing, the latter of which could equally be driven by convection from heating at the pole as at the equator.[57]

Fultz's results were first published in a unique report commonly known as the '*Staff Members*' paper.[58] Journal articles are typically referred to by the authors' names and the date, as in 'Hadley (1735)' for example, but this one in contrast is simply attributed to the Staff Members of the Department of Meteorology of the University of Chicago. Only inside the paper is a list of names hidden: an all-star cast of now-famous meteorologists, with Rossby as the leader.

The scope of the paper is daunting, covering many different aspects of atmospheric circulation, but a central theme is the newly-discovered jet stream. In this paper, perhaps for the first time, the jet is truly recognized as a global feature: 'the middle latitude belt of westerlies generally may be traced as a continuous stream around the globe but that the belt itself is surprisingly narrow and may be described as a meandering river winding its way eastward through relatively stagnant air masses to the north and south'.

The Staff Members' paper highlights a special quality that meteorology possesses as a relatively young science. In other areas, scientists can often be classified by type, for example as a theorist who works exclusively with equations, or an experimentalist, who might work in a team of thousands on some gigantic device designed to smash tiny particles together. In contrast, even today in meteorology it is common for a scientist to work partly in two, or even all three, of the fundamental areas of observation, theory and modelling. The Staff Members paper leads with the observations: our first glimpses of the scale, shape and structure of the jet stream above us. Then follows the mathematical theory, largely due to Rossby, of how the jet could be formed by the mixing of vorticity on a planetary scale in the atmosphere. Finally, there is the 'verification' by experiment: Fultz's results showing that a jet can indeed be formed from just the two ingredients of rotation and mixing.

The jet stream was now a major topic of study in universities around the world. The Second World War, as other conflicts, had spurred on new developments and discoveries in meteorology, and the beginnings of a global observing system were taking shape. By the time of the first Soviet atomic weapons test in 1949, American meteorologists were able to estimate the location of the test site by tracking the jet backwards over Asia from the West Pacific, where radioactive particles had been found by US planes flying out of Alaska and Guam.[59]

Rossby had correctly identified vorticity mixing as a crucial element of jet formation. (Vorticity, if you remember, is a measure of how much a small bit of fluid is spinning.) But his theory was lacking in one key regard: it hadn't taken on board the inherent instability of the mid-latitude atmosphere. He missed the role of the storm track in transporting heat poleward and hence spreading the jet winds down to the ground. This was first identified independently by two great meteorologists on either side of the Atlantic. One was Jule Charney in Los Angeles, who would go on to be instrumental in the first successful attempt at operational weather forecasting by computer.[60] The other was a remarkably independent Englishman by the name of Eric Eady

Like Fultz, Eady had been a meteorological consultant during the Second World War, working on upper air analysis and forecasting for bombing and reconnaissance missions. His last posting was in Lagos, Nigeria, just a short trip along the ITCZ from Freetown. What spare time he had, he devoted to filling notebooks with scribbled equations in pursuit of his dream: the construction of a satisfying theory of atmospheric instability. He returned from the war to Imperial College, London, where he soon gained his PhD and a job on the faculty.

Eady's masterpiece is a pen-and-paper theory for the growth of mid-latitude storms. There were no powerful computers behind him, just the equations of atmospheric fluid dynamics, through which he blazed a pioneering path. His form of the equations predicted not only that storms would grow, but also what their rate of growth and their shape would be, including the all-important westward tilt with height. Finally, the theory showed that the jet would be fundamentally changed by the storm, with the upper winds weakened and the lower winds strengthened.

Eady possessed a rare genius but left relatively few scientific papers,[61] a result of his singular personality. Quoting his obituary in the *Quarterly Journal* of the Royal Meteorological Society, 'he could handle well a motor bicycle and a sports car; he had a keen appreciation of nature, music, literature and art, and talking with him was one of life's pleasures. But in his attitude to his work he was a perfectionist grappling with phenomena of great intrinsic difficulty and complexity...'.

Eady's theory remains one of the few major results which can be derived purely by playing with the equations in wartime notebooks. Beyond this, the complexity of atmospheric fluid dynamics withholds simple answers. As Eady himself declared, 'any theory of the atmospheric circulation must be based on a theory of (large-scale) atmospheric turbulence'. Turbulent fluid flow is messy and chaotic: think of the billows in a plume of smoke, or the breakers crashing onto the beach at Soup Bowl. It is 'the most important unsolved problem of classical physics', in the considered opinion of the great physicist Richard Feynman. Eady, as a result, grew increasingly frustrated and depressed by the sheer difficulty of his subject. He still had great ambitions but, in his own words, he finally realized that 'he was not going to be able to build a cathedral'. He died in 1966, at the age of fifty, from an overdose of painkillers.[62]

It would be another decade after Eady's death before computing power reached the level where it could really help to discover what lay in the realm beyond his theory. By the 1970s, however, novel experiments were beginning to be performed not only in the lab, but also on the computer. A classic set of experiments was executed in England, at the University of Reading, by Brian Hoskins and Adrian Simmons.

Eady's model of storms had been *linear*. This means that all the really complicated terms in the equations of motion had been ignored. The assumption was that we only care about small deviations of the flow from its average state. Any term which involves multiplying two of these small deviations together, for example, would be so small that it could be ignored. This simplification allowed Eady to solve his equations on paper, but to retain these terms and solve the full equations would need a powerful computer.

Simmons and Hoskins used a state-of-the-art computer at a national facility in London. State-of-the-art in the 1970s was, of course, slightly different from today, and their runs were at the limit of what was feasible at the time. To use less computer memory they imposed symmetry on their problem, with their growing storm repeating itself six times around the Earth, as if it was trapped in a house of mirrors at a funfair. Even so, their most detailed simulation had to be slipped onto the computer on Christmas Eve, when no-one else was using it. Then came a nerve-racking wait to see if the run had been successful; around a week later they received a late Christmas present when the results arrived on micro-film, courtesy of the Royal Mail.

The design of their experiment was simple. They set the model up with a very unstable initial state: a strong jet restricted to upper levels, just as the jet looked to Grantley when he emerged from Asia into the Pacific. Then they simply let it run forward in time, like a weather forecast. In effect, they pushed the rice pile up to be artificially steep, and then let go. In Eady's version the landslides of rice down the side of the pile were relatively small events, just correcting the slope a little. But in the computer simulations, more dramatic things were to happen.

In the first few days of the run, the system evolved much as Eady had predicted, with cyclones and anticyclones growing in a chain around the mid-latitudes. But as the weather systems grew larger, their structure began to alter, influenced more by the complicated, nonlinear terms in the equations. As hoped, these changes brought the storms into closer agreement with those in the real atmosphere. The systems strengthened at upper levels in particular, and started to develop characteristic southwest-northeast tilts in their shape as shown in Fig. 8.2. Down at ground level, sharp warm and cold fronts developed at the boundaries between warring air masses.

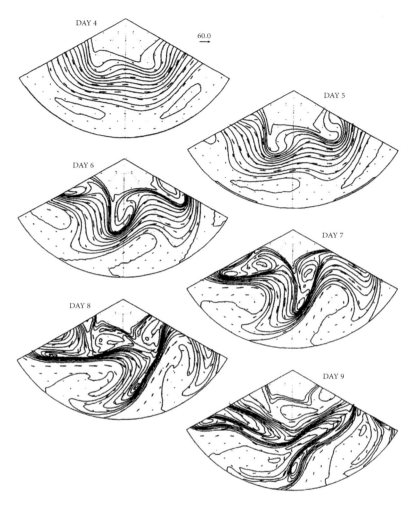

Fig. 8.2. Day-by-day development of the lifecycle experiment, showing the change from a gentle wave pattern into a deformed structure which tilts from southwest to northeast.

After a week of simulated time, the systems stopped growing and began to weaken. After another week they had faded away to jumbled ripples in the flow, mere debris as remnants of the great storms. This process of storm decay was something entirely new; nothing like it could occur in the Eady model. Hence, these experiments are known simply as the *lifecycle* experiments.[63]

The lifecycle experiments have revealed many secrets that the atmosphere had previously withheld. Most importantly for our story, they showed how the jet stream is profoundly altered by the decay phase of storms. In the Eady model, storms just re-arrange the jet winds, spreading them downward in the atmosphere so that the jet weakens above as it strengthens below. But by the end of the lifecycle experiment, in contrast, the jet has actually become stronger overall. Despite the growth of such violent weather systems, and the ultimate reduction of the north-south temperature contrast, the westerly winds up at Grantley's level end up even stronger than they were initially.

The crucial ingredient in this overall driving of the jet is the tilted shape of the systems. The southwest-northeast lean that develops in the flow is perfectly designed to pump momentum into the jet from the subtropical side, as the tilted cyclones thin out and stretch down to lower latitudes.[64] The lifecycle of these computer storms shows us how energy cycles through the atmosphere. Initially there is potential energy, since light and dense air masses sit side by side, but this is converted into kinetic energy of swirling storms. Then, as the storms decay, this energy is put to use in accelerating the jet. This energy cycle is named after Edward Lorenz, the pioneering meteorologist who deduced its basic structure back in the 1950s, using only the limited observations of the day and a heap of mathematical and physical insight.[65]

In explaining the jet stream we have now used arguments about momentum and also arguments about energy. Can we just pick and choose these as we see fit? Is the jet best understood as driven by flows of momentum or by cycles of energy? Elmar Reiter neatly compared this to two museum visitors admiring a statue, one from the front and one from the back. While both ultimately admire the same statue and derive the same basic meaning from it, the details they have each actually seen are different. Energy and momentum conservation are both fundamental concepts in physics, so both *have* to hold for the jet. But by circling round the statue and admiring it from both angles we get the deepest possible understanding of it.[66]

Reiter, as an aside, was pre-eminent among a generation of 'jet stream meteorologists' in the 1950s and 60s. He wrote two books on the topic;

Fig. 8.3. Multiple jet streams indicated by the parallel cloud bands of Jupiter.

an authoritative textbook entitled 'Jet Stream Meteorology' and also a more approachable student's companion text. Both are wonderful detective stories recounting the challenge of his day, which centred around developing cunning new ways to collect upper air data in order to trace out the jet stream and reveal its deepest secrets.[67]

The lifecycle experiment is just one example of how computer simulations of the fluid dynamics equations can give great insights. These models are also perfect for playing God, enabling us to ask incredible 'what if' questions. For example, what would happen if we took a more complete model of the atmosphere, with storms growing and decaying all the time, and gradually increased the size of the planet?[68] At normal

Earth size this model would produce a jet encircling the globe which would be a merged jet, as we have found over the Pacific, arising from both subtropical and eddy driving.

As the planet gets bigger, however, the eddy-driven jet would shift poleward and ultimately separate out from the subtropical one. At this point, there would be two distinct jets in each hemisphere. Moving from the equator to the pole, we would pass through the Hadley cell to the subtropical jet, and then beyond this would be a clear gap before the eddy-driven jet would be reached. The separated jet would be purely eddy-driven, accelerated only by the momentum moved around by those tilted weather systems.

If we pushed on, and made the planet larger still, a third jet would appear, and then a fourth, and so on. For a large planet, much of the atmosphere would be filled with a whole array of parallel jet streams. Each of these would be a deep, eddy-driven jet, accompanied by its own storm track.

In fact, we do not need to turn to computer models to see this effect. Close-up pictures of the gas giants Jupiter, Saturn, Uranus and Neptune all reveal striking banded structures encircling the planets (Fig. 8.3). These indicate the presence of multiple jets, made visible through differences in cloud cover. These observations support the implication from the computer models: if the Earth was just a bit bigger, or spun a little faster, it would support not one merged jet, but two separate ones.

As Grantley passed over Japan we learnt how Ooishi took the first systematic measurements of the jet stream. But was he the first jet stream observer in history? Perhaps this prize should have been awarded much earlier, although its recipient didn't really know what he'd seen. It was in the 1660s, many decades before even Hadley's theory, when Giovanni Cassini peered through his telescope into the clear Bologna sky and first glimpsed the bands of Jupiter.

Simulating multiple jets in the laboratory required engineering beyond that available to Dave Fultz. One particularly clear demonstration was performed in 2014 by a team led by Peter Read of Oxford University. They used the 'Coriolis Platform' in Grenoble, a huge fourteen-metre turntable which could carry a tank longer than a bus, filled with a hundred tonnes of water. The tank was set spinning for three to four days at a time, at its fastest completing a full circle every forty seconds. The tank had a sloping bottom, designed to mimic the spherical shape of the Earth,[69] and several kilometres of electrical cables underneath provided gentle heating to trigger convection and hence drive the water into motion. The resulting flow was highly chaotic, an indoor sea of swirling eddies and swaying Rossby waves. Excitingly, the system formed not one, but several west-to-east jet streams flowing around the circular tank. These were not constant, stationary features, like the bands of Jupiter, but instead fluctuated erratically: meandering, merging and splitting from hour to hour.

A remarkable conclusion, both of these lab experiments and of many computer simulations, is that the spacing between jets roughly agrees with something termed the *Rhines scale*.[70] This length scale therefore determines how many separate jets will fit in the tank. It is named after Peter Rhines, the Seattle-based oceanographer who is equally at home tinkering with tanks in the lab, equations on the board, and robotic 'sea-gliders' used to explore the real ocean. Rhines' theory is based on turbulence, as Eady had said that any theory should be. The recipe is simple: turbulence plus Rossby waves equals jet streams.

Turbulence in this case involves things growing, as small vortices merge together to form larger ones.[71] In Rhines' words, however, this 'cascade of pure turbulence to large scales is defeated by wave propagation'. As the vortices get bigger, they feel Earth's rotation more and eventually trigger Rossby waves, which then quickly propagate away. The Rhines scale is simply the length scale at which the Rossby waves take over. The waves largely propagate in the west-east direction, along the jets, hence the Rhines scale at which the cascade halts becomes the typical scale of separation between the jets.

The main message of this chapter is that jets invariably form in rotating fluid systems, be they on planets, in tanks, or encoded on a computer. These jets don't need to have a Hadley cell attached to them, they are most often purely eddy-driven. It is just the size of Earth and its rotation rate that mean Grantley is now following a single, merged jet across the Pacific.[72]

As Rossby realized, for jets to form you just need mixing and rotation. Rhines formalized this by using the mathematics of turbulence theory to describe the mixing, but it is the Rossby waves which are all-important. While the waves often follow the jet, the crucial process which actually accelerates it is the eventual propagation of Rossby waves *out* of the jet. It is the wave leaving the jet, angled towards the equator, which gives the vital southwest-northeast tilt to the eddies. As in the decay phase of the lifecycle experiments, it is this tilted structure which pumps momentum into the jet, making it stronger overall as well as deeper: a truly eddy-driven jet stream.[73]

To end this discussion, we return to the tiny island of Guam in the tropical Pacific. Situated firmly under the Hadley cell, Guam lies well to the south of Grantley's jet. Yet despite this, the island is in fact surrounded by jets: unseen westerly and easterly currents tracking their way across this giant ocean. Flow features such as eddies and currents are much smaller in the ocean than in the atmosphere, due to the very different density structure of the fluid.[74] Hence the Rhines scale is much smaller, and while Earth's atmosphere can support just one or two jets, the ocean can fit tens of jets into the space between equator and pole.

While Dave Fultz pored over some of the early measurements of the atmospheric jet from wartime Guam, he couldn't have known that the nearby ocean was swimming with similar jets, a whole array of them stacked one after the other from Guam all the way north to Kamchatka. He presumably also had little suspicion that he would go on to pioneer the field of laboratory experiments for atmospheres and oceans, or that supersized descendants of his classic experiments would eventually replicate this sea of transitory jet streams, meandering, splitting and merging all around him.

Even as his renown spread, and problem after problem succumbed to his equations, Joseph was never truly pleased with his own work. He grew to be fiercely self-critical, always claiming much more appreciation for the work of others than of himself. Upon completing a new theorem he would most likely tear it up, only to start again the very next day. He was constantly undoing and redoing his work, and even then was often only passably satisfied with the result. He would write earnestly to Leonhard and others, imploring them to suppress any of his work which they considered to be of lesser value, for fear that it might embarrass him. And as each year passed he became harder to please with the quality of his own work.

Niño

Just a few days after leaving Japan, Grantley has made it across the largest ocean on Earth and glides in high over the American west coast. But where exactly? Up to Japan, his trajectory was clear; he could have been launched in 2019, 1920 or any year in between, and he would have followed a very similar route. Even during the Russian heatwave of 2010, the jet would have only swung him north by a few degrees of latitude. The subtropical jet is notable for its reliable steadiness, just like the trade winds that it is linked to through the Hadley cell.

But now that we have hit the storm track and the jet has become at least partly eddy-driven, it is much more variable. The eddying weather systems that drive the jet here are turbulent and chaotic, and so its path can change significantly from year to year, or even from week to week.

Grantley's point of arrival in America is therefore quite uncertain (see Fig. 9.1). In the winter of 1998/99, for example, he would have likely hit the coast just south of the Canadian border, near Mount Baker in Washington State. Storm after storm followed the same path that winter, helping the Baker ski resort secure the world annual snowfall record with a remarkable 29 m falling in that one season.

Eleven years later the jet would have taken him much further south, and just over the border from Mount Baker the organizers of the Vancouver 2010 Winter Olympics had to resort to trucking in snow.

In December 1964, Grantley would have arrived in the US over the Oregon/California border, witnessing the devastating 'Christmas floods' at first hand. Several were killed in the two states in a deadly combination of cold then wet; Portland Airport, for example, already had eleven inches of snow on the ground when the rains began.

Finally, had Grantley made landfall in January 1958, he would have travelled way south, cruising in over Los Angeles in southern California.

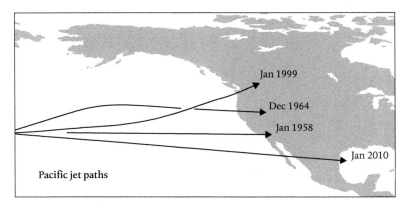

Fig. 9.1. Examples of the path taken by the jet stream approaching North America.

Hollywood would have been thriving below him, with Sinatra's *Come Fly with Me* ruling the charts and the first stars in the Hollywood Walk of Fame laid that very year. Surf shacks were springing up along the coast, although the community they served was still very much a fringe culture. Yet here again the flood waters were rising and the southward-shifted jet was causing disruption from LA right across the country to a distant storm-hit Florida.

The unfolding events of January 1958 were being examined at first hand by a keen eye from an office in UCLA (The University of California at Los Angeles). Forty years after he had changed the history of meteorology by poring over Norwegian weather maps, Jack Bjerknes was now a professor at UCLA. He had come to the US on an eight-month lecture tour at the onset of the Second World War and stayed ever since. Once again, he was poring over maps, but crucially this time he was looking at patterns of ocean temperatures as well as airflows.

Sometimes the jet can be displaced or disturbed for no particular reason other than the random fluctuations of the weather systems. But in this region, in particular, the strongest jet shifts are often orchestrated by an unseen conductor, and this was the puzzle that Bjerknes began to unravel in 1958. This was the International Geophysical Year, in which a global push was made to capture and record data from around the world, and in a lucky coincidence a great global weather event was unfolding.

The Peruvian coast is in many ways similar to that of Nouakchott: ocean upwelling provides cold, nutrient-rich waters off the coast, while inland lie the bleak subtropical deserts that were branded the lands of the 'tearless skies' by Herman Melville in Moby Dick. Peruvian fishermen for generations had been familiar with a warm coastal current which regularly disrupted the upwelling, arriving every year around Christmas and hence earning the name *El Niño*, or Christ Child.

While normally benign, there were sporadic years when the warm current would be much stronger than usual. The results were often devastating: fish stocks would collapse and islands off the coast would be littered with the bodies of dead sea birds. The tearless skies would weep and the desert would flood. Such events have very likely struck this region for centuries, and have been postulated to explain the location of ancient Inca settlements unusually far from rivers.

One of the major revelations that struck Bjerknes in the wake of the 1958 El Niño is that the warm waters were not restricted to the Peruvian coast, but incredibly they spread out along the equator into the very middle of the Pacific. This was revealed in vital ocean observations taken during the International Geophysical Year. The tiny atoll of Canton Island in the equatorial mid-Pacific, roughly halfway between Hawaii and Fiji, was a key data point. In January 1958, the sea temperature here was found to be unusually warm, just as it was off Peru six thousand miles to the east.

Over the coming decade, Bjerknes was to compare this ocean data with wind measurements and some of the earliest satellite photos to form a bold new theory. Crucially, he linked the ocean warming to a change in something called the *Southern Oscillation*, which was a somewhat daring move to make at the end of his career. The Southern Oscillation was a brainchild of Sir Gilbert Walker, who had claimed that weather patterns far around the world were actually connected to each other. Many, however, were not convinced by these links, and another leading meteorologist recalled thinking of Bjerknes at the time: 'that poor old man, he's gotten tangled up with Walker. His reputation will be ruined.'[75]

Gilbert Walker was an accomplished applied mathematician at Cambridge; he graduated in 1889 as the Senior Wrangler (i.e. the best maths student of his year) and soon won a lectureship. His obituary would describe him as 'modest, kindly, liberal-minded, wide of interest and a very perfect gentleman', and his interests were certainly wide, ranging from flute design and the aerodynamics of bird's wings to skating and gliding. He was fascinated by the boomerang, both in theory and in the field, so to say, and was hence nicknamed 'Boomerang' Walker at Cambridge.

In 1904, Walker left Cambridge to take up the role of Director-General of Observatories in India, and was soon immersed in trying to develop a mathematical way to predict the Indian monsoon. The monsoon rains provide vital fresh water and also act as a pressure valve to alleviate the intense summer heat. When the monsoon fails, the consequences are extreme, for example in the devastating summers of 1877 and 1878 when the resulting famines are thought to have killed over 15 million people across India and China.

In his new role, Walker had a veritable army of civil servants at his disposal, and access to decades of weather measurements from across the British Empire. As well as the monsoon failure, 1877-78 had also been noted for unusually high pressure over Australia and anomalous weather around the world, such as drought in Brazil. Seeming coincidences such as these motivated Walker to search for links between weather records from around the globe, with surprising results.

Walker's main findings, published in the early 1920s,[76] uncovered three great see-saws in atmospheric pressure. Two of these were located over the northern oceans (one in the Pacific and one in the Atlantic) and we will return to these later. The third described a 'swaying' of pressure across the tropics, between the Indian and Pacific Oceans, and it was this that Walker christened the Southern Oscillation.

The Southern Oscillation is most commonly measured by the pressure difference between Tahiti in the mid-Pacific and Darwin in northern Australia, as these locations are the most strikingly anti-correlated: when the pressure is higher than usual at one of these points, it is very likely to be lower than usual at the other. In most years Tahiti has the higher pressure, but occasionally the pattern can reverse. Walker noticed that this reversal often coincided with the monsoon failures, and so used this as the basis of his monsoon forecasts.

Walker died in 1958, the very year that triggered Bjerknes' interest in El Niño. His methods had failed to predict some monsoon busts and so his world weather correlations were falling out of favour. But like one of his precious boomerangs, Walker's ideas were about to come hurtling back to take centre stage in the story of El Niño.

Bjerknes realized that the ocean warming of El Niño was very closely related to the reversals of the Southern Oscillation index. The two are in fact so closely related that their names are often merged into *ENSO*, short for El Niño Southern Oscillation. This highlights the truly coupled nature of the phenomenon: take either the atmosphere or the ocean out of the picture and the whole thing would fall apart.

The crucial mechanism, now termed the *Bjerknes feedback*, was postulated in his 1969 paper. The warmest sea water is usually found in the so-called 'warm pool' in the westernmost tropical Pacific, kept there by the action of the trade winds blowing from the east. But during El Niño the trade winds weaken and the warm water spreads eastwards across the Pacific.[77] Whichever region of sea is warmest becomes a factory for thunderstorms and hence deep convective ascent of air up through the atmosphere. As the warm water spreads east, so does the region of convection, and soon Canton Island is surrounded by warmth and soaked in rain. As the convection moves, the trades weaken further and so the warming strengthens; the *feedback loop* is completed.

The link between the convection and the trade winds was a central part of Bjerknes' theory, and one that builds directly on Walker's work. Walker had suspected that some type of airflow linked his high and low pressure regions, and Bjerknes connected this to the all-important changes in the trade winds. It was Bjerknes that named this great air current the *Walker Circulation*.

The Walker Circulation is essentially a sideways version of the Hadley cell. Air rises up over the warmest parts of the tropical surface and then outwards at upper levels, but instead of moving poleward, this flow moves zonally, i.e. in the east-west direction along the equator. The air then sinks back down in the relatively cooler regions of the tropics. In the tropical Pacific the usual Walker cell comprises air rising in the west and moving east at high altitude. The return flow from east to west at the surface makes the trade winds particularly strong there.

The ascending region of the cell is where the low pressure is to be found, along with the storms, and the descending region is the dry, high pressure zone. Hence, the see-saw in pressure between the two regions is fundamentally linked to fluctuations in the strength of the Walker Circulation. In El Niño conditions, the Walker cell over the Pacific is weakened and displaced and the trade winds are weakened as a result (see Fig. 9.2).

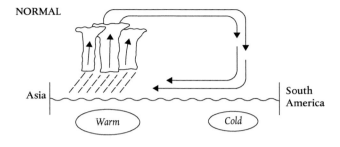

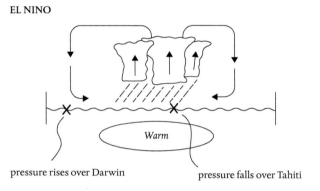

Fig. 9.2. Changes in convection and the Walker Circulation that accompany ocean warming during El Niño.

El Niño also has a flip-side; a little sister called *La Niña*. In a La Niña year, roughly opposite behaviour occurs: the Walker cell and the trade winds are both stronger than usual, enhancing the contrast between cool water in the east Pacific and warm water in the west. The behaviour of both siblings is wrapped up in the term ENSO.

Key to our story is the effect of El Niño on the jet stream; this is the invisible hand which pulled the jet south and steered all the storms towards Bjerknes as he studied charts in LA in the winter of 1958. To some extent we can understand this as another Rossby wave pattern: the enhanced convection in the central and eastern Pacific during El Niño triggers a Rossby wave that arches north towards Alaska then bends south again over North America. This wave features a large cyclone to the west of North America, which weakens the westerly wind approaching the northern parts of America and strengthens that approaching the south.

However, there is also a broader effect on the jet, which reaches to some extent around the globe. By allowing warm water to spread across the tropical Pacific, El Niño slightly, but noticeably, warms the whole of the tropical atmosphere. The temperature contrast between the tropics and mid-latitudes is therefore strengthened, and by thermal wind balance the subtropical jet is also strengthened.

Now arises one of the major complexities of the climate system. Recall that the jet is effectively a merger of two different entities: one driven by the Hadley cell and one by the eddies. The subtropical, i.e. Hadley-driven, part of the jet has been altered by El Niño, but this change also affects the eddies; the cyclones and anticyclones which both propagate on the jet and also drive it. By changing the subtropical winds, El Niño has altered the patterns of how these eddies grow, move, distort and decay. Once the eddy behaviour has been changed, it is inevitable that the eddy-driven part of the jet will be changed somehow as well.

In fluid dynamics, this complex behaviour is generally known as *eddy-mean flow interaction*. It is a type of feedback, just as the Bjerknes feedback is. In the Bjerknes case, the feedback involves the Walker circulation affecting the ocean temperatures, which affect the Walker circulation, and so on. In this case, the jet affects the eddies, which affects the jet, and so on....

One of the major challenges in meteorology is that it is very hard to predict from theory alone how the eddies will affect the mean flow, which in this case is the jet. In general, some kind of computer model is needed for this step. The feedback problem is so complex that it can't be solved with pen and paper, which is what so frustrated Eady and many others after him. In the case of El Niño, the end effect is that the eddies, and hence also the eddy-driven component of the jet, shift slightly closer to the equator.[78] In this way, the whole depth of the jet is shifted southward from Oregon to California during El Niños, not just the upper level, subtropical flow. And the 'eddies', of course, are the same mid-latitude cyclones that battered both LA and Florida in the winter of 1958.

By many measures El Niño represents the greatest phenomenon in natural climate variability. In the space of a year it reorganizes the whole global Walker circulation, displacing the regions where air rises and falls within the tropics. By shifting the jets and by sending anomalous Rossby waves around the globe it exerts a profound influence on regions far from the equatorial Pacific.

As well as the Indian monsoon failures and the Australian floods and droughts noted by Walker, El Niño has also been blamed for flood failure of the Nile, mudslides in Mexico, and drought in Brazil. It affects the maize yield in southern Africa, the sugar crop in Hawaii and the orange farms of Florida, and has set the scene for devastating wildfires in Java and Borneo. It has influenced the spread of malaria in South America and various epidemics in central Africa. It shifts the global pattern of hurricane activity and has been implicated in major coral bleaching events. It likely even played some part in the distant Russian heatwave of 2010.[79]

With such a global array of impacts, El Niño is understandably a focus for the science of *seasonal prediction*. Just as Walker's goal was to provide early warning of catastrophic monsoon busts, weather forecasting centres around the world strive to predict the occurrence and impact of each El Niño event in advance. Based on this, and other signals, the leading centres routinely issue forecasts for the average conditions of the season to come; will this winter be colder or wetter than usual, for example, or bring increased risk of windstorms?

The ocean's involvement in El Niño proves a real advantage here, as the sea conditions evolve much more slowly than weather patterns, which can often change completely in a matter of days. By November we generally have a pretty good picture of how the ocean temperatures will look for the coming winter, and this offers great potential for predicting the likely atmospheric conditions. El Niño is just one of the potential sources of information for seasonal prediction, with others ranging from sea ice changes and large volcanoes to regular variations in the radiation emitted by the Sun, but it is by far the most important.

By May 1997, for example, the Pacific sea surface temperatures were rising rapidly and it was clear that a major El Niño event was on the way. Based on experience of previous events, scientists at the US National Oceanic and Atmospheric Administration (NOAA) Climate Prediction Center predicted very wet conditions in California for the coming winter, including even the possibility of deadly mudslides; predictions which sadly proved correct. Although they share the same underlying dynamics, not all El Niño events play out in exactly the same way. These predictions were successful because they were based not on all past

El Niño events, but only the strongest ones, such as the one Bjerknes experienced in 1958.[80]

Fast forward to 2015 and another strong El Niño was developing. Eyes all around the world were focused on the equatorial Pacific, as this event was the most eagerly anticipated for at least two reasons.

Firstly, a large El Niño would likely lead the world out of the so-called slowdown in global warming. The temperature variations associated with these events are so large that they appear in the record of the Earth's global average surface temperature. The last strong El Niño had been the one in 1997/98, and this had warmed the planet globally by a fraction of a degree. In the years following this bump it seemed that the pace of global warming had slowed. Hence, the public debate over climate change was reignited and many chose to believe that warming had stopped altogether. The reality is that the warming continues but natural variations such as El Niño mean that some decades warm particularly fast and others less so.

Secondly, a wet and stormy winter in California would have been widely welcomed for once. Like the Sahel, California straddles the boundary between two climatic zones, and so has a very variable climate. When the jet is shifted north, the state lies in the dry, stable subtropics, but when the jet moves south it can get all the mid-latitude storms going. California is hence susceptible to both floods and droughts, the former largely striking in El Niño years and the latter more sporadically, with notable examples in the 20s, 30s and 70s.

By spring 2014, California was suffering from the effects of three unusually dry winters in a row. While not unprecedented, this was enough to cause major disruption. To the residents of LA and the other urban centres, it caused only minor irritations such as short showers and brown lawns. However, California is the USA's largest agricultural producer, and in this regard things were not looking so good. In 2014 alone the drought is estimated to have cost California $2.2 billion and 17,000 agricultural jobs. While many farms were able to offset the damage, this was done by pumping up groundwater at an unsustainable rate, leading to subsidence and water shortages elsewhere. Other impacts included low air quality and enhanced wildfire activity, which unusually even extended into winter.

The California drought was typical of many prolonged extreme events in that it was not, to the best of our knowledge, caused by any one factor.

Each of the three dry winters was slightly different, though all involved a northward jet shift linked to anticyclonic weather patterns off the California coast. (In unusually colourful language, scientists dubbed this the Ridiculously Resilient Ridge.) This setup seems to have been at least encouraged by a pattern of Pacific Ocean temperatures at the time (distinct from El Niño), and the resulting atmospheric Rossby waves, but this influence doesn't explain it all. The severity of the drought was likely enhanced simply by the random variability of weather patterns.[81]

So the excitement mounted through 2015 as it became clear that the developing El Niño would indeed be one of the strongest on record. Would it deliver? Certainly, the 2015/16 El Niño provided a dramatic burst to end the 'slowdown' in global warming, with 2015 and 2016 both coming in almost half a degree warmer than 1998. A few years on and global temperatures have remained high. The slowdown was an interesting example of climate variability that was about more than just El Niño, but looking back at the full record it was just a minor blip; a false summit on the path to ever increasing global temperatures.[82]

For California, however, relief was not to come at the hands of the 'Christ Child'. It wasn't until 2017, after six years of drought and following the record wildfire summer of 2016, that the jet finally shifted back south and the heavens opened of their own accord. Not all El Niños have exactly the same patterns, both in terms of ocean temperatures and atmospheric waves, and this event just didn't work out for California. While El Niño generally disrupts weather patterns around the globe, and is the cornerstone of seasonal forecasting, it doesn't always play by the book.

This example serves to illustrate the intrinsic uncertainty of climate variability. The jet stream can be influenced by distinct, remote factors such as the Walker circulation over the tropical Pacific. The driving by these factors is indeed often a useful source of skill for seasonal forecasts, but only in the sense that it loads the climate dice and alters the probabilities.

For example, El Niño loads the dice in favour of a wet and stormy winter for California, but it doesn't make that a certainty. You can still throw a one even with a dice that's loaded in favour of a six, just less often. For the jet there can always be other effects at play, perhaps competing drivers or perhaps just the wonderfully unpredictable randomness of the atmosphere. Sometimes the jet just does its own thing and that's that. Despite the El Niño, the jet was displaced north around much of the Northern Hemisphere in the winter of 2015/16, and so California stayed dry. In an unlikely hemispheric symmetry, it was hoped that the same El Niño event would bust Cape Town out of its own multi-year drought, and again it failed.

Sometimes when established patterns break down, this just reveals another level of complexity. Gilbert Walker's results were perhaps not taken seriously enough because his monsoon forecasts ultimately achieved mixed success. The iconic monsoon failure of 1877/78 exhibited the telltale pressure reversal between Darwin and Tahiti, but other cases did not, and hence came without warning (the more recent 2002 and 2004 seasons are good examples of this). It now seems that a weak monsoon can occur in non-Niño years, but it is indeed more likely during El Niño events, especially those with the strongest warming located in the central longitudes of the Pacific.[83]

The successful seasonal forecasts for California in 1997 were examples of *statistical forecasts*, based on past data and experience from previous events. By 2015, the *dynamical forecasts* obtained from computer models were much more advanced and the best of them could outperform the statistical models in many cases.[84]

The dynamical forecasts are generated by computer models which are very similar, if not identical, to weather forecast models. They step forward in time using the fluid dynamics equations to predict future weather patterns. In contrast to weather forecasts, they are run for several months rather than days ahead, not to predict each individual storm of the winter but to predict the statistics, to see how the climate dice have been loaded that winter by El Niño and by whatever else is going on.

The dynamical models have one essential advantage over the statistical models. Being based on the principles of physics rather than on past experience, they are potentially much better at making so-called 'out-of-sample' predictions, i.e. predictions of things which could in theory happen but haven't yet, at least as far back as our records go. These models really do have the potential to predict the future, rather than just repeat the past, and are vital stepping stones to the models used to provide early warning of future climate change.

Joseph was widely noted for his distinctly reserved nature. Quiet and unobtrusive, what many took for kindness was better interpreted as extreme passivity. Quite simply, what mattered most to him was his equations, and he avoided all attempts to be pulled away from them. He steadfastly avoided conflict, even when this meant abiding by rules he considered unfair. Even when others criticized his work he would resist being drawn into quarrels, responding instead with kind words and praise for their insight, before carrying on with his work regardless. While deeply modest and sweet of temper, he could be obstinate in sticking to his principles.

CHAPTER 10

Mountains

Had Grantley been launched in the winter of 2015/16, he would have arrived in the US a long way north, despite the ongoing El Niño. Chances are he would have hit the Olympic coast in Washington State, just south of the Canadian border. This part of his journey would have been a bumpy ride, with large storm systems spinning him round and veering him off course. Thankfully, most of the bubbling storm clouds and associated rainfall would have been far below him.

But suddenly, as he approached the coast, the air around him would have got busier. After his long, lonely Pacific crossing he might have met another balloon, a fellow radiosonde sent up from below to measure the atmosphere. He also risked an impact from above in the shape of a *dropsonde*: a bundle of instruments similar to a radiosonde but dropped from above, so that it can sample the atmosphere on the way down. He might even have met instrument-laden research planes on strange zig-zag flight paths, the very planes which were dropping the dropsondes.

The reason for all this activity is that Grantley would have drifted straight through the middle of the OLYMPEX field campaign,[85] a NASA-funded project to blitz the area with a whole array of different observing systems. New weather satellites orbiting high above the Earth can detect precipitation (i.e. rainfall, snow, etc.) better than ever before but they still need 'ground-truthing', by comparing their data to that from *in-situ* measurements such as from planes and balloons. A particular need was for good data on rainfall over mountains, and for this the scientists had definitely come to the right place.

Just a few kilometres inland, the ground below Grantley would start to rise up towards the Olympic mountains. Peaking at just under 2500 m, these are mere foothills compared to the ranges to come, but to the moist air flooding in off the Pacific this is more than enough height. As the air is forced to rise it expands, cools and releases its cargo of moisture.

Ranked in many datasets as the wettest place in the lower forty-eight, these hillsides are hence home to some of America's best remaining temperate rainforest. The giant spruce, hemlock and Douglas firs are almost perpetually dripping with water during the wet season, and hundreds of species of moss, lichen and ferns coat every available surface. The National Park Service website proudly claims that 'even the air seems green here'.

The OLYMPEX project certainly got enough rain and snowfall to test the satellites with. The northward-shifted jet graced the region with storm after storm that winter, providing the scientists with several excellent case studies to help understand how mid-latitude cyclones and fronts behave when they hit a mountain range. As well as planes, balloons and dropsondes, the forest was peppered with ground-based observations from simple rain gauges up to truck-mounted radar units. Equipment had to be carried in to some of the remotest stations by mule trains slogging over the forested slopes (in the rain of course). This is perhaps the only time that NASA has had to resort to mules to deploy its instruments.

The data from these ground stations testifies to the power of the mountains. Stations on the beaches recorded 1700-1900 mm of rain over the winter, while just 50 km inland a station in the Quinault river valley received an incredible 4900 mm. That's much more rain than Freetown, located right under the ITCZ, gets in a year. Mountain ranges have an innate ability to wring moisture out of the atmosphere. On the lee side of the Olympic and nearby Cascade mountain ranges, for example, lies the so-called Canadian desert: the Okanagan Valley which does get enough rain to sustain fruit orchards and even vineyards, but nowhere near enough for a rainforest.

In addition, mountains can have a profound effect on the jet stream, and nowhere is this effect stronger than over the great Rocky Mountains. The Rockies, at around 4000 m, are only half as high as the Himalayas yet despite this they have a stronger impact on the jet stream.[86]

One difference is that while the Himalayas run roughly from west to east, in the same direction as the jet, the Rockies are aligned south to north. These mountains hence form a great barrier to the westerly winds all the way from the desert lands of New Mexico up to the snowy forests of British Columbia. Another difference comes from the jet itself. When Grantley passed the Himalayas, the jet was purely subtropical in nature and so only a high-altitude feature. Now, as he approaches the Rockies, Grantley is riding an eddy-driven jet which extends all the way down to the ground. So not only is there a better barrier in the form of the Rockies, there is also a much deeper jet stream approaching them.

When this deep jet hits the Rockies, the air near the surface is forced to rise, climbing up the mountain sides. In a simple conceptual picture of this flow we can think of how this interaction affects a *column* of air: let's pick out a vertical tube of air stretching up from the ground and follow this as it hits the mountains. This is a simplification of course, as the column would be moving faster at upper levels where the jet is stronger, so would soon tip over, but it is still a useful guide to what happens.

When the base of this air column is pushed upwards by the mountains, this acts to squash the column from below. To see what effect this has, we return to the ice skater analogy. Squashing the column makes it fatter as well as shorter. This is like the skater extending their arms to spread their weight, and this causes them to spin more slowly. For our atmospheric column, this means that its vorticity has decreased.

Let's assume that the column had zero relative vorticity to start with, which means that it wasn't spinning with respect to the ground. It still had planetary vorticity though, as it was spinning along with the Earth. Then, as the column is squashed, its total vorticity decreases, so it is now spinning less than the Earth. From the ground, then, it appears to spin the opposite way to the Earth, i.e. anticyclonically. Soon, however, the column passes over the top of the mountain range and gets stretched back to its normal form as it rides down the lee side (Fig. 10.1).

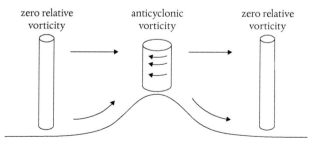

Fig. 10.1. A column of air gets squashed as it flows over a hill, creating an anticylcone over the hill.

This doesn't just happen to our column, of course, but also to all the others that follow it. The result is a large, near constant anticyclone sitting right over the mountains, as each column in turn gets squashed. As Grantley approaches the Rockies then, he veers north a bit, following the flow clockwise around this giant anticyclone. The jet is ultimately a stronger feature than the anticyclone, however. He doesn't get sucked in to the eddy to loop round and round it like in a whirlpool, but he does get diverted north by a thousand kilometres or so before swinging back south on the downwind side of the mountains to rejoin his eastward path.

This, however, is not the end of the story as far as the mountains go. We have created a Rockies-sized anticyclone and this has a strong effect on the air around it, triggering a giant Rossby wave train downstream of the mountains. Similar to the Rossby waves we met over Asia, this anti-cyclone induces a cyclone to its east, which induces another anticyclone even further east, and so on. This wave doesn't go straight east though, due to the spherical shape of the Earth, but curves south towards the equator. The snaking of the jet around this curving wave train will have profound consequences for Grantley when he finally makes it back over the Atlantic Ocean, as we will soon see.

The Rockies therefore affect the atmospheric circulation far down-stream to the east by setting up what is known as a *stationary wave* pattern.[87] This is a slight misnomer, as nothing is ever completely stationary in the atmosphere. The anticyclone over the Rockies is constantly heaving and swaying, as is the downstream cyclone. Some-times these features are not even there at all, when large travelling Rossby waves or weather systems distort the flow. But they are there so often that they clearly exist *on average*, since the mountains are always there and always exerting the same force on air columns. If you take the wind data and average it over a month, say, you will almost always see the jet swing north around an anticyclone over the Rockies and then south again around a cyclone, as shown in Fig. 10.2.

So, for as long as the Rocky Mountains have been there, they have acted to divert the jet. Now is a good moment, with Grantley drifting high over the Rockies, to pause and briefly consider the distant past. Earth's climate has varied incredibly over the planet's lifetime, with the early atmosphere likely made from a completely different mix of gases to today.[88] Even over the 'recent' period, measured in millions rather than billions of years,

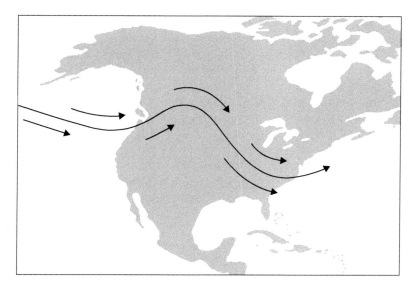

Fig. 10.2. The typical displacement of the jet stream by the Rockies anticyclone and the Rossby wave downstream.

there have been dramatic changes in global temperature and ice cover. But what about the jet stream? How different was that? How far back do we have to wind the clock before we reach a climate where the jet looked really different from today?

The answer, sadly, is that we don't really know. After all, the jet proved so elusive that it wasn't discovered until the twentieth century. It will always remain hidden behind the scenes in the otherwise valuable records we have of past climates, such as sediment or ice cores, which only tell us about surface quantities like temperature. But the physics-based computer models can be useful here as well, and some of these suggest that the atmospheric circulation could have been radically different as recently as 50 million years ago.[89]

The key characteristic that would be needed for this is extreme heat. Sea surface temperatures in the tropics today average around 27°C, and for serious changes to the jet these would have to be ramped up to 35°C or even higher. Achieving this in the computer models requires twenty to thirty times the present-day amount of carbon dioxide, a change many times larger than the one we are effecting now.

Such climates are understandably termed 'hothouse' climates. The latest of these were the in Eocene, from 34 to 56 million years ago, and the Cretaceous period, from 66 to 145 million years ago. These hothouses were also what are termed *equable* climates, with a much lower equator to pole temperature contrast than today. In some models of these periods, the tropics are so hot that they practically bubble with atmospheric convection, storm cells continuously popping up and merging with others into giant organized cells. This drives an entirely different type of circulation, with legions of Rossby waves forced in the tropics that propagate outwards towards the poles. Just as the present day jet is sustained by Rossby waves propagating *out of it*, this activity also drives a strong westerly jet. But it is not located at the latitudes of Soup Bowl, Tsukuba and the Swat river valley. Instead it is centred directly over the equator, where the skies bubble most fiercely.

It is possible then (but by no means certain), that in these past climates the Earth had just one jet, positioned bang over the equator, as shown in Fig. 10.3. This situation is remarkably odd. Remember that the subtropical jet that carried Grantley all along the first half of his journey was driven by air moving poleward from the equator. This air acquired an eastward velocity over the ground, since nowhere else on the spinning Earth was moving as fast as the equator; the air simply left

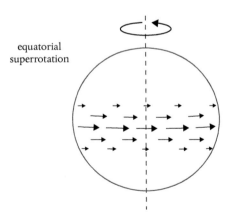

Fig. 10.3. A superrotating atmosphere, with just one jet stream running along the equator.

the slower-moving ground behind. In the purported hothouse climates, Rossby waves pumped so much momentum into the tropics that the air over the equator was actually moving faster than the Earth itself, hence forming a westerly jet. Our much-used ice skater analogy has finally been broken here, as this corresponds to the skater stretching out her arms and yet somehow ending up spinning even faster.

Such a state is termed *superrotation*, as the tropical atmosphere spins faster than the Earth underneath it. This state can never be achieved simply by moving air masses across the latitudes. Implausible as it sounds, evidence for this type of atmospheric circulation has been found on Venus, Saturn, Jupiter and Titan (the largest moon of Saturn). So it is possible that, during these very hot periods in Earth's past, the planet had a strikingly different jet stream from today.

To put this in context, it was during the Cretaceous that the super-continent of Pangea completed its break up, so the world as a whole looked very different. Dinosaurs still roamed the land, for example. By the Eocene, a few million years later, the continents were approaching their current form but the all-important Rocky Mountains were just beginning to rise. It is a testament to the robustness of the jet stream that we have to go back this far to have even the possibility of a radically different jet from today's. As Dave Fultz learned in his lab in Chicago, all you need is rotation and mixing, and the jet will form.

There have, however, been more subtle changes to the jet since then. If we zoom in to just the most recent 2.5 million years of Earth's history, we discover the ice ages. The continents were essentially laid out as they are today, crucially with large areas of land over, or near, both poles. This greatly facilitated the development of ice caps during cold periods, which is otherwise hindered by the warm and fluid ocean. The ice ages themselves were triggered by slight variations in Earth's orbit and developed due to feedback processes, such as the albedo feedback in which snow and ice reflects more solar energy back to space, hence cooling the planet further.

This is known as the Quaternary Period, itself made up of shifts between glacial and interglacial periods. One of the most studied of past climates is that of the last ice age, specifically the *Last Glacial Maximum* a mere 21,000 years ago. Things were subtly different compared to today, in terms of Earth's orbit and the balance of gases in the atmosphere.

But several experiments have suggested that these factors weren't so important for the jet. What mattered more was the ice itself.

At the height of the Last Glacial Maximum, the eastern half of North America was covered by the giant Laurentide Ice Sheet, which rose up to two miles high at its peak. It extended south as far as present day New York and Chicago, and was at least partly responsible for gouging out the basins that now form the Great Lakes.

Crucially for the jet stream, the ice sheet acted to 'fill in' the lowland American continent east of the Rockies. It was the elevated surface, rather than the cold itself, which was most important. The jet effectively encountered a whole continent of high elevation, rather than just the relatively narrow Rocky mountain range. This forced an even larger mountain anticyclone and significantly changed the shape of the downstream Rossby wave train. So the change in the jet stream was not global but regional, only apparent in an alteration to the path of the jet after it passed North America and crossed the Atlantic Ocean.[90] But we are getting ahead of ourselves now, the Atlantic twist in the tale of the jet stream is yet to come.

We should first return to Grantley and let him complete his crossing of America. His trajectory curves back southward as he leaves the Rockies behind. Here over the middle of North America the jet can often strengthen again, as Grantley's eddy-driven flow combines with some other airflows: a smaller portion of the jet which snuck south around the Rockies and a new subtropical westerly air stream also coming in from further south.

A local intensification of the jet such as this is termed a *jet streak*, and these are often closely monitored. It turns out that there are two key points along a jet streak which are particularly prone to storm growth. These are the *right-entrance* and *left-exit* regions, i.e. on Grantley's right-hand side as he accelerates into the jet streak and on his left-hand side as he slows again on the way out. The flow in these places encourages air to ascend underneath the jet which helps convective storms to form.[91] In winter, jet streaks will often be eagerly watched by North American skiers anticipating a fresh dump of snow. In other seasons they are monitored more nervously, as the large convective storms associated with a jet streak can often spawn devastating tornadoes.

Even without these effects, the wintertime jet over North America brings plenty of severe weather. After its diversion around the Rockies, the flow is now heading south-east, leaving much of Canada and the northern American states with bitterly cold prevailing north-westerlies. This southeastward-flowing air has the Rockies anticyclone to its west, and to its east is a large, semi-permanent cyclone: the next part of the stationary wave pattern that the jet snakes around. The jet loosely comprises the boundary between mild air to the south and cold air to the north, so when it veers north as it does over the Rockies this is effectively a protrusion of warm mid-latitude air towards the Arctic. In contrast, as the jet next flows south around the cyclone it encloses an extrusion of cold, polar air bulging down over eastern Canada.

The inhabitants of the northeast corner of North America are therefore well prepared for cold conditions, but even so the winter of 2013/14 took many of them by surprise. In this year the cyclone extended further south than usual, blanketing much of eastern North America with frigid Arctic air from the northern side of the jet. Ice and snow dominated the whole winter for many, with several separate waves of cold air spilling south over America. At its height, 92% of the area of the Great Lakes froze that winter, the highest percentage since 1979.

One cold surge peaked on 7 January 2014, when several media outlets declared that all fifty states experienced temperatures below freezing at some point. Even Hawaii qualified, although only due to the cold experienced at the top of a very high volcano. Back on the mainland, sub-freezing temperatures reached as far south as central Florida. A little further north in Kentucky, a hapless convict successfully managed to escape from a minimum-security prison only to turn himself in again several hours later to escape the cold. Further north still, staff at Chicago's Lincoln Park Zoo made the headlines when they decided to keep Anana, their polar bear, inside in the warm for the day.

Such widespread cold, however, was not typical even for this winter. While the eastern half of the continent shivered in what was the coldest season since the seventies for many, the west was actually unseasonably mild on average. This, after all, was one of the winters that made up the California drought. The proximate cause of all this was an amplification of the stationary wave pattern: the Rockies anticyclone was stronger than usual and so was the downstream cyclone, with the jet snaking further up into the Arctic than normal and then plunging back down to the deep south.

The prolonged cold was taken by some as the final proof they needed that global warming was no more than a hoax. John Holdren, the Chief Scientific Advisor to the Obama White House, issued a rapid response to this. Holdren made the surprising claim that these events might actually have been helped, not hindered, by the recent warming, stating 'I believe the odds are that we can expect as a result of global warming to see more of this pattern of extreme cold in the mid-latitudes and some extreme warmth in the far north.'

Holdren was endorsing a recent theory that centred on the Arctic. As the pattern of climate change has emerged, it has become clear that the Arctic is warming faster than the rest of the globe, since effects such as the melting of sea ice can locally enhance the warming. This had been predicted by climate models years before, but the pace of polar warming even seemed to be outstripping many of their projections. As we learnt in Chapter 4, the jet stream is in thermal wind balance with the temperature contrast between northern and southern latitudes. So as the Arctic warms and the equator to pole temperature contrast weakens we might expect the jet to weaken. The theory is then that the weakened jet is more prone to taking large meanders and getting stuck in amplified Rossby wave patterns like the one that spread the cold unusually far south in the 2013/14 winter.[92]

This is certainly an interesting idea, and it is quite possible that as the jet stream responds to climate change, the array of Rossby waves that propagate along it might be affected. However, many climate scientists are unconvinced by the specifics of the Arctic theory. As concluded by one high-profile opinion piece which urged caution in the wake of Holdren's comments, the idea deserves a fair hearing but the arguments so far are not very compelling.[93]

Part of the challenge is that enhanced Arctic warming has only become apparent over the last couple of decades at most. But mid-latitude weather patterns can vary strongly from one year to the next for no apparent reason, so detecting systematic changes against this high level of background noise generally requires longer time periods than this.

The task is made even harder by other sources of variability between decades, such as the variations in Atlantic temperatures that were so important in the story of Sahel drought in Chapter 4.

Hence, the observational data ultimately proves inconclusive. For hypotheses such as this to become accepted, they also require support from theory and from computer models to demonstrate causality; it would need to be shown that Arctic warming is actually *causing* changes in Rossby wave behaviour, not just that they happen to be observed at the same time.

So for now at least, there is not strong evidence that the rapidly warming Arctic is already seriously affecting mid-latitude weather patterns, though clearly it will at some point in the not too distant future.[94] Despite this, the Arctic hypothesis has been supported so widely and frequently in the media that it has entered the public consciousness, for better or for worse.

So what actually caused the extreme winter of 2013/14, if it was not primarily the Arctic? It is possible, in fact, that there was no real 'cause' to speak of. Atmospheric weather patterns are complicated and often unpredictable, and it is quite possible for something like this to just happen for no apparent reason. Pure atmospheric variability should really be the null hypothesis, or default theory, to explain any weather event.

In this case, however, there is some evidence that a different remote agent likely did play a role in the 2013/14 winter. The world-wide weather upheaval associated with El Niño demonstrates the clear potential of the tropics to affect the jet stream. Rossby waves are much more readily triggered from the tropics than from higher latitudes, as the kind of vigorous ascending updrafts that get high up enough to disturb the jet core are much more common in the warm tropics.

That winter, the sea was unusually warm in the tropical West Pacific, but further west than the region where El Niño rules, and there were strong updrafts in the atmosphere above. It doesn't seem to explain the whole story, but some computer simulations have suggested that it was this warmth, not the Arctic, that helped to enhance the usual stationary Rossby wave train over North America, maintaining the California drought and bringing severe cold to the eastern continent.[95]

Despite his outwardly reserved aspect, Joseph's mind raced and surged within, like a river in spate. While working on a problem he would still often reach childlike levels of excitement. But just occasionally this giddiness would develop into something else. Then the symbols would start to dance on the page before him, and he would be forced to retire, cursing the interruption. The cause was an irregular pulse, an affliction from which he was to suffer throughout his life.

CHAPTER 11

Gulf

 The morning of 5 September, 1862 was damp and misty in
Wolverhampton, England, yet excitement there was mounting.
James Glaisher and Henry Coxwell were busying themselves with
preparations as their giant balloon slowly inflated, hooked up to the
local gasworks. In their few ascents so far, the pair had made it up to
around 24,000 feet, just above the long-standing altitude record set by
the French pioneer Gay-Lussac. This time, they were determined to go
higher.

At three minutes past one in the afternoon, they cast off the lines and
rose up into the clouds. Shortly afterwards they burst out into brilliant
sunshine, and were rewarded with a stunning scene. 'Beneath us lay a
magnificent sea of clouds, its surface varied with endless hills, hillocks
and mountain chains, and with many snow-white tufts rising from it.'[96]

Little time was taken to admire the view, however, as Glaisher
systematically attended to each of his instruments in turn, recording
the detailed changes in the atmosphere as Coxwell guided the balloon
smoothly up through its layers. Intently focused on their work, the
sky turned darker blue around them, and Glaisher's readings tracked
the temperature dropping steadily. The cold soon began to bite, even
though the pair were well wrapped in heavy coats and scarves.

Glaisher's instruments started to frost up and fail, but still they
threw out ballast and pushed on upward. Hunched over his equipment,
Glaisher struggled to make out the mercury in the thermometer and
to read off the small marks on the scale, and he realized his eyesight
was going. Looking upward, he called out to Coxwell for help but was
met by an alarming sight. The balloon had been spinning as it rose,
and as a consequence the valve line used to release gas had become
tangled. Without this, there was no way to halt the balloon's perilous
ascent, and to fix it Coxwell had climbed up out of the basket and was

perched precariously in the rigging. The barometer suggested the balloon had reached a height of 29,000 feet, and was still rising fast.

Glaisher's condition, meanwhile, was declining rapidly. He realized that he was no longer able to move his own arms, and then his head began to flop around uncontrollably. Moments later he lost all vision and collapsed against the wall of the basket, before losing consciousness altogether.

He was revived some minutes later by Coxwell, who had somehow made it safely back into the basket after releasing the valve line. He had needed to tug at the line using his teeth, as his hands were too frozen to grip it. Ever the professional, Coxwell was now urging Glashier awake with cries of 'temperature', 'observation', and 'do try; now do'! Glaisher was just able to respond, declaring 'I have been insensible'. Coxwell replied 'You have, and I too very nearly', and showed Glaisher his hands which had turned fully black from the cold.

As the balloon dropped down, Glaisher was soon recovered enough to administer some Victorian medicine, by splashing brandy on Coxwell's hands. Then he recommenced his measurements at seven minutes past two, just ten minutes after passing out, recording a temperature of $-19\,^{\circ}$C.

Glaisher claimed to have suffered no 'inconvenience' following his trauma, and indeed the pair managed to walk seven miles to the nearest village after landing in a field in the Shropshire Hills. The same, however, cannot be said of the six pigeons which they had taken with them in the balloon to release on the way up, by way of experiment. Glaisher noted with interest that one fell rather like a piece of paper, while another simply dropped like a stone. On landing, one was found dead at the bottom of the basket and another somewhat dazed. Two days later just one of the six, for some reason, decided to return to Wolverhampton.

Glaisher later estimated their maximum height to have been 37,000 feet, which at over 11 km is clear above the summit of Everest and could well have put them in the stratosphere. This trip firmly established them as celebrity explorer-scientists,[97] and a widely published lithograph from the time dramatically shows 'Mr Glaisher insensible at the height of seven miles' (see Fig. 11.1).

The pair went on to make many more ascents over the following years, although were subsequently careful to stay below 24,000 feet. The measurements taken by them and other 'aeronauts' provided

MR. GLAISHER INSENSIBLE AT THE HEIGHT OF SEVEN MILES.

Fig. 11.1. Mr Glaisher insensible at a height of seven miles.

invaluable data for understanding how the atmosphere varies with height, including early insights into the high altitude wind patterns. On several occasions, balloonists were caught out by the strength of the wind higher up and were forced to make emergency landings to avoid being carried out to sea. In 1864, after one of their less eventful trips, Glaisher noted the presence of a warm, south-westerly wind, in a comment which proved to be well over a hundred years ahead of its time:

> The meeting with this SW current is of the highest importance, for it goes far to explain why England possesses a winter temperature so much higher than is due to our northern latitudes. Our high winter temperature has hitherto been mostly referred to the influence of the Gulf Stream. Without doubting the influence of this natural agent, it is necessary to add the effect of a parallel atmospheric current to the oceanic current coming from the same regions – a true aerial Gulf Stream.

Just a few years later, in 1868, the French scientist Gaston Tissandier made a similarly prophetic statement after his maiden balloon flight, during which interleaving air currents had carried him out over the North Sea from Calais and then safely back again:

> Who can say that aeronauts will not some day discover a true system of circulation in the atmosphere, with its various veins and arteries, its regular and periodical currents, and its Gulf Stream, the course of which a balloon will be able to follow as surely as a sailing vessel glides over the surface of the mighty ocean?

It was, of course, not manned balloons that finally charted this system of veins and arteries, but Rossby's fleet of automated radiosondes. It is time we returned to our own balloon, surely following the course of a great stream over the surface of the Earth. Grantley has made it safely across North America and, somewhat by coincidence, is just about to pass over the Gulf Stream.

Even in Glaisher's and Tissandier's day the Gulf Stream was well known and credited for maintaining the mild climate of western Europe. But still today, confusion often arises between this and the jet stream due to the similar terminology. Put simply, one is a flow of water, the other of air. The Gulf Stream is a local ocean current, in which warm seawater flows northward along the Atlantic coast of North America. The jet stream, of course, is a global atmospheric phenomenon; an air current that can carry an unsuspecting weather balloon from the Caribbean right around the world.

The stationary wave in the lee of the Rocky Mountains took Grantley south-east past the Great Lakes and now out over the East Coast, passing just south of Washington D.C. The warm current of the Gulf Stream snakes through the ocean immediately to the south of his path, having separated away from the US coast at Cape Hatteras.

As he drifts high over the coast, Grantley might well feel a tinge of *déjà vu*. He has passed over a great continent and met a southwest-northeast oriented coastline. The ocean is warmer than the land in winter, so this orientation means the usual south-to-north temperature contrast is enhanced here. Place on top of that a particularly warm and narrow ocean current, and the temperature contrast is strengthened further. Just like the Kuroshio off Japan, the Gulf Stream is the icing on the cake that makes this region one of the world's leading cyclone factories. It was here, in 1991, that the so-called 'Perfect Storm' formed, which went on to inspire a best-selling book and film. This was an extreme example of the mid-latitude cyclones which spawn here prodigiously, although somewhat atypical in terms of track and behaviour compared to many.

The earliest recorded encounter with the Gulf Stream was in 1513, when the Spaniard Juan Ponce de León sailed north from Puerto Rico and became the first European to set foot in Florida. On turning southward to survey the coast, the explorers were startled to find their ships actually moving backwards with respect to the land, and hence they deduced the presence of a great current surging northward.

One of the most popular early theories was that this was simply the outflow of the great Mississippi River, bursting out into the Atlantic from the Gulf of Mexico.[98] Draining water from thirty-one different states, the Mississippi is certainly a giant among rivers. However, simple calculations show that this is only one thousandth of the amount of water carried north in the Gulf Stream.

The Gulf Stream, like the Kuroshio, is an example of a *western boundary current*, the narrow return flow which is an intrinsic part of the great ocean gyres. As such, it is the driving of the ocean by the prevailing wind patterns which is responsible, and central to the story is our old friend Ekman.

Briefly, Ekman's theory tells us that the water under the mid-latitude westerlies should actually move southward due to the three-way balance between wind stress, Coriolis and pressure. To support this flow, water rises up to the surface on the northern side of the wind belt and sinks down on the southern side. This vertical motion then affects the whole column of seawater below, right down to the ocean floor.

Columns of water on the southern side, for example, get squashed by the water pushing down from above. This squashing would imply a decrease in vorticity through the ice skater effect. Recall that the vorticity is made up of two parts, the planetary component due to Earth's rotation, and the relative component due to the fluid spinning relative to the Earth. In the ocean, the planetary component of vorticity dominates over the relative component, so the easiest way for the vorticity to decrease is for the whole column of water to drift south, to lower latitudes where the planetary vorticity is weaker.

In this way, it is the whole depth of the ocean that is moved by the winds, not just the surface waters. The result is a steady drift away from the wind belt, to the north on the north side and to the south on the south side. Homing in on the subtropical gyre then, we see that this is hugely skewed: we have southward flow all across the basin from Florida to Morocco, but this must return north somehow, hence there must be an incredible northward current flowing back, clinging to one or other of the coasts.

It was Henry Stommel in 1948 who first developed a mathematical model of this by including the all-important frictional terms in the equations. He found that a strong return flow on the western boundary of the ocean (i.e. the American side) could not only balance the budget of sea water, but also that of vorticity. All across the subtropical gyre, the anticyclonic wind pattern around the subtropical high acts to pump anticyclonic vorticity into the ocean. In Stommel's model this vorticity can only be removed by friction acting on a strong boundary current on the western edge of the ocean (see Fig. 11.2). This is because the current there is flowing north, so the friction of the coast drags on the left-hand side of this current, imparting an anticlockwise (i.e. cyclonic) spin to the water there.

So the Gulf Stream baffled sailors for four centuries before finally giving up its ultimate secret. Many, however, learnt the intricacies of its flow in practical terms, such as American merchant ships and whalers

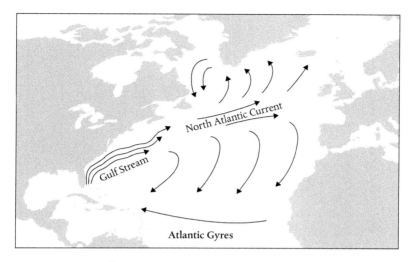

Fig. 11.2. Horizontal ocean currents in the North Atlantic, showing southward flow across the subtropical gyre and the narrow return current which is the Gulf Stream.

who would cruise along the edge of the stream in search of whales. It was to a Nantucket sea captain, Timothy Folger, that Benjamin Franklin turned for help in 1769. Franklin was at the time working in London as Deputy Postmaster-General of the Colonies, and was frustrated that the mail ships to America often took up to two weeks longer to reach their destination than merchant ships such as Folger's. Folger knew that the British ships were unaware that they were sailing against the Gulf Stream, for having tried to warn them on several occasions he found that 'the English captains were too wise to be counselled by simple American fishermen'.

Folger promptly sketched out the path of the current for Franklin, who had it drawn up as the first ever printed chart of the Gulf Stream (reproduced in Fig. 11.3). This pioneering map is reasonably accurate, showing a strong current squeezing through the gap between Florida and the Bahamas and then clinging to the American coast northward before separating and heading out east towards Europe.

Some versions of the map also feature a strange circular eddy around the whole Atlantic basin. This is blatantly not realistic, but then it appears

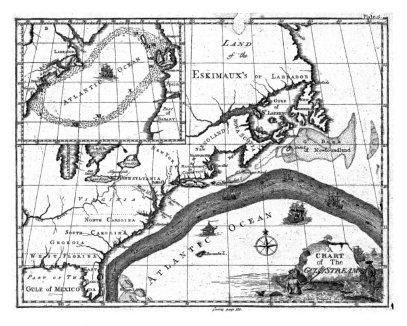

Fig. 11.3. Franklin's map of the Gulf Stream, with the herring migration route inset, where it should be.

this was never part of Folger and Franklin's original map. To save money, the American version of the map included as an inset another map of the Atlantic from an entirely different paper by John Gilpin, illustrating his new theory for herring migration routes. Unfortunately, the two maps were combined into one in some reproductions, and Gilpin's herring appear as a mysterious pan-Atlantic current.[99]

Another hundred years on, and the Gulf Stream had been championed by Matthew Maury in his 1855 book *The Physical Geography of the Sea*. Maury achieved great things, in particular as one of the first to make detailed charts showing the typical wind strength and direction over the oceans, by painstakingly compiling information from ship's logs. Unfortunately, in the words of Henry Stommel who finally unlocked the secret of the Gulf Stream, 'I cannot suppress the view that Maury was very much confused concerning fluid mechanics'.

Maury was even considered a fraud by many in the scientific community of his time, despite his organizational skills in cataloging the ocean winds. Several commentators have suggested that his greatest contribution was in writing such implausible theories in his book that they motivated the 'shy genius' of William Ferrel to turn his attentions to meteorology.[100]

It was Maury's book that popularized the idea that western Europe owes its mild winter climate to the Gulf Stream:

> 'One of the benign offices of the Gulf Stream is to convey heat from the Gulf of Mexico, where otherwise it would become excessive, and to disperse it in regions beyond the Atlantic for the amelioration of the climates of the British Isles and all of Western Europe. [... otherwise] the soft climates of both France and England would be as that of Labrador, severe in the extreme, and ice-bound.'
>
> Matthew Maury, The Physical Geography of the Sea, 1855

This remains a widely held belief to this day, with many a local tourist office on the western Scottish Isles proudly advertising their few, somewhat weather-beaten, palm trees as evidence of the warming influence of the Gulf Stream. There are, however, a few problems with this claim.

Firstly, the Gulf Stream doesn't really come anywhere near Scotland. It runs north along the American coast before splitting off in the direction of Europe, but then it meanders and breaks up into a multitude of swirling eddies. The water off Scotland is indeed surprisingly warm for its latitude, but this is due to something different which is often called the North Atlantic Current.

Although this seems something of a technicality, it reflects the fact that very different processes are at work. While the Gulf Stream is a feature of the ocean's gyre circulation, in which water moves largely horizontally around the basin, the North Atlantic Current is more associated with the vertical *overturning circulation*. This comprises a net flow of warm water northward in the upper layers of the ocean which, when it reaches the Arctic, is cooled so much that it becomes relatively dense and so sinks down to lower levels. The southward return flow therefore happens at depth, underneath the northward flow.

It is the overturning circulation which transports warm water up to the latitude of Scotland and beyond, not the Gulf Stream. So, when worries are voiced that the Gulf Stream might shut down under climate change, it is really the overturning circulation which is meant. As part of the horizontal gyre system, the Gulf Stream is driven by the winds blowing across the Atlantic. Hence, oceanographers like to say that the Gulf Stream will not shut down so long as the Sun keeps shining and the planet keeps spinning.

However, a shutdown of the overturning circulation is indeed possible, although generally considered unlikely. This was vividly portrayed in the blockbuster 'The Day After Tomorrow', with ice sheets and killer storms buffeting New York. This scenario will be discussed in more detail later on, but for now be reassured that the specific events in the film are about as likely as those in Jurassic Park.

Regardless of terminology, is it really the warm seas of the North Atlantic that nurture the Scottish palm trees? Surprisingly, this question was not properly addressed until 2002, in a controversial paper led by Richard Seager of Columbia University.[101] Few in the climate science community had really thought the European warmth was all about the ocean, but still the results of this paper put many oceanographers on the back foot.

Following Maury, Seager focused on the difference between western Europe and Labrador, at the same latitude as the British Isles but directly across the Atlantic on the east coast of Canada. Remarkably, the British Isles are typically $15-20\,^\circ$C warmer than Labrador in winter, and Seager's team analysed observed data and climate models to work out why. Somewhat surprisingly, the ocean turned out to play a minor role in this, only warming Europe by a few degrees.

Just as Glaisher had speculated all those years ago, Scotland's palm trees largely owe their existence to the great 'aerial Gulf Stream'; the very current that is carrying Grantley on his journey around the world. On leaving North America, the course of the jet stream takes a pronounced northward turn, such that it doesn't head straight east towards the Mediterranean, but instead aims north-east, directly for Scotland. This explains the prevailing south-westerly winds that many in western Europe are used to, as the jet is largely eddy-driven in this region and so extends right down to graze the Earth's surface. Since the wind blows from a southerly direction, it brings air from lower, warmer latitudes, and this goes a long way to explaining the unusual winter warmth.

So does the ocean do nothing to help? Clearly if there were no ocean here at all then Europe's climate would be more continental than maritime, with larger seasonal swings between colder winters and warmer summers. And the overturning circulation does indeed warm Europe; when this circulation is artificially inhibited in computer simulations then Europe cools by 2 or 3°C. This cooling, however, also spreads over Labrador and beyond. While clearly important for both Labrador and Europe, it doesn't therefore explain the difference between the two regions, which is what Maury and Seager both focused on. What it does explain is much of the difference between Scotland and southernmost Alaska, also at the same latitude, since the overturning circulation which warms the North Atlantic does not have an equivalent in the North Pacific.

To understand the Scotland-Labrador issue, we therefore need to explain the path of the jet stream as it crosses the Atlantic. This was the aim of a novel set of computer simulations performed by David Brayshaw and colleagues at the University of Reading. They figured that the detailed shape of the coastlines and mountain ranges shouldn't matter much, and tried to see if they could understand the structure of the jet and the Atlantic storm track from more basic building blocks.[102]

Instead of North America, they put into their model a simple, triangular continent, with the narrow end pointing south where Mexico is in reality. All the complexity of the topography, with ranges of jagged peaks and deep canyons, was replaced with one smooth ridge running south to north to represent the Rocky Mountains. Just these two ingredients, it turned out, were enough to explain a large fraction of the essential northward tilt of the jet as it heads across the Atlantic towards Europe.

Just as Grantley found in our story, the jet was forced to swerve around a stationary Rossby wave train forced by the mountain ridge. It largely went north around the Rockies anticyclone and then plunged back south over the continent, passing around the southern side of the downwind cyclone. Beyond this cyclone, the jet swung back northward, but by this point it had reached the edge of the peculiarly angular continent. So the model generated a northward tilted jet, just as in reality, as part of the Rossby wave which appeared downstream of the mountains.

It is this stationary wave which particularly enhances the temperature contrast between Labrador and Europe. Just past the Rockies, the jet is angled southward and so provides the prevailing north-westerly winds which keep the interior of North America so cold. Labrador, on the eastern edge of this continent, feels this cold influence profoundly. But then the next meander in the Rossby wave points the jet back north, bringing a south-westerly influence and warm air to Europe. So what makes Scotland so much warmer than Canada? It is largely the Rocky Mountains.

In addition, more subtle effects also emerged in Brayshaw's simulations. The mountains made the jet tilt from southwest to northeast after the stationary cyclone, and this happened to align more or less with the eastern coastline of their great American triangle. For Grantley, riding the real-life jet stream over the actual continent, the same thing happens. He finds himself heading northeast towards Europe, roughly running along the American east coast which happens to point in the same direction. Importantly, this situation enhances the potential for storm growth, as the jet aligns parallel to the strong temperature contrast between the bitterly cold land on Grantley's left and the much warmer ocean on his right. So storms that grow here end up running along the coast, feeding for longer off the energy in the land-sea temperature contrast.

As if this wasn't enough, one more effect emerged in the simulations. Brayshaw added some equally rough details to the Atlantic Ocean in his model. He formed a tight Gulf Stream along the coast, bringing warmer water right up to the edge of the frigid land, and finally a large warm region further north, representative of the effect of the ocean overturning. The response to these was certainly subtler than the response to adding the mountains, but it proved to be the icing on the cake, extending the line of the warm-cold temperature contrast out towards Europe and tilting the jet that little bit more towards the north. So while the ocean seems to have a lesser effect than the mountains, it does indeed help to keep Europe a little bit warmer than it should be, given its latitude, partly just by its warmth but also by helping to shape the path of the jet stream across the Atlantic.

So several ingredients come together to make the Atlantic part of the jet stream special, with a larger deviation from the west-east direction than anywhere else along Grantley's path. They also form one of the greatest mid-latitude storm factories on Earth, where the Rocky mountain wave sets up a flow right along the coastline which separates cold, dry Arctic air from the warm, moist air over the Gulf Stream. In many ways, this is the ultimate 'Perfect Storm'.

Aware that he was suffering increasingly from overwork, Joseph retreated into a rigidly regular routine. Every morning he would rise late and diligently address his correspondence until lunch. Then he would take one of his long, solitary walks, breathing deep in the open air and drifting off into rambling meditations. Then he would regimentally shut himself up in his room from six until midnight to work alone, and it was from these evenings of self-imprisonment that his great theories emerged. Isolated and protected by his uniform life, the days and the years passed by. Joseph grew ever more distant from the family and friends around him but burned increasingly brightly among his companions of letters.

CHAPTER 12

Split

Let's return to Grantley, currently making his way across the Atlantic towards Europe. It's a bumpy ride again, thanks to all those storms. What's more, something important has happened to him, as you might have noticed. Like Phileas Fogg, Grantley has successfully made it the whole way around the world! And what's more, he's managed it in about about eight days; only a tenth of the time taken by the hero in Jules Verne's classic adventure story.

By imagining that Grantley has stuck steadfastly in the centre of the jet, we've also helped him to beat Bertrand Piccard and Brian Jones, who took seventeen days to make the first non-stop round the world balloon flight in the Breitling Orbiter 3 in 1999. While largely following the jet, weather patterns forced them to make some tricky manoeuvres along the way in order to keep it as a tailwind as much as possible.

After crossing the US east coast, Grantley followed the jet northeast towards Scotland, paralleling for a while the Gulf Stream ocean current far below him. Just a fifth of the way across the Atlantic he has reached the 59°W meridian. Barbados lies on this very line, but crucially it's about 2000 miles to the south. Despite making it the whole way around the world, Grantley has not actually ended up back where he started.

As we saw in Chapter 11, the layout of mountains, coastlines and ocean currents has conspired to tilt the jet northeast across the North Atlantic. Hence, the jet stream does not form a complete ring around the Northern Hemisphere, but instead spirals inward towards the Arctic. Grantley has been drifting imperceptibly northward for much of his journey, gaining a few degrees of latitude over Africa, a couple more across Asia and more still by the time he reached the Gulf Stream. But now he makes a major turn to the north, and between here and Europe he will drift north by as much again as on the whole of the rest of his journey. Such is the strength of the jet tilt across the North Atlantic. This is the source of the south-westerly winds that, as well as warming Europe, returned Columbus and many other ships from the Caribbean to Europe in the years of the infamous triangular trade.

The jet tilt, as an aside, is quite seasonal. In summer the Hadley cell is much weaker, and hence the whole jet is weaker and is also shifted north by several degrees. If Grantley had been launched from Barbados in summer, rather than winter, he may still have made it into the jet stream but it is less likely. If he did indeed make it into the jet, his subsequent trip would have been slower and also simpler, as the summer jet is much more of a straightforward ring around the globe.

Back in winter, however, the spiral path reveals a fundamental change in the nature of the jet. Across the North Atlantic the two types of jet, subtropical and eddy-driven, have become split. When Grantley started his journey, the jet was purely subtropical (i.e. driven only by the air moving north at high altitude in the Hadley cell). Now, as he veers strongly north away from the Hadley cell, his jet has for the first time become purely eddy-driven.[103] This means that the only mechanism pushing the southwesterly winds into northern Europe is the action of the fleeting eddies and weather systems which continually pump momentum into the jet. As we mentioned in Chapter 8, if the planet was just a little bigger then the jets would separate out and there would be two parallel jets encircling each hemisphere. In this relatively small sector of the Atlantic, thanks to the stationary Rossby wave downstream of North America, this splitting of the jets has actually been achieved on Earth.[104]

Driven purely by the turbulent mid-latitude eddies, the jet across the North Atlantic is much more variable than if it were driven by the steady action of the Hadley cell. We have described its path in somewhat average conditions, but some years over the North Atlantic are far from average. Here the jet can shift north and south more than anywhere else in the hemisphere, with dramatic effects on weather patterns around the Atlantic basin and beyond.

In one extreme, the eddy-driven jet can shift so far south that it actually merges with the subtropical jet, as shown in Fig. 12.1. At such a time, Grantley would be steered toward the Mediterranean rather than Scotland, and could in fact find himself sucked back into the subtropical jet over Africa to undertake another lap around the world. Once established, this situation can persist for several weeks, starving northern Europe of the warming effect of Glaisher's aerial Gulf Stream. In the winter of

Jet north (e.g. 2011/12) - clear jet split

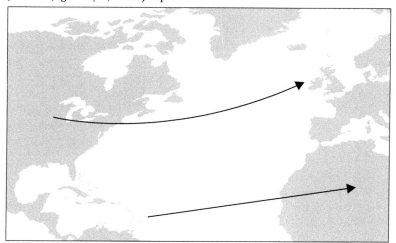

Jet south (e.g. 2009/10) - jets merged

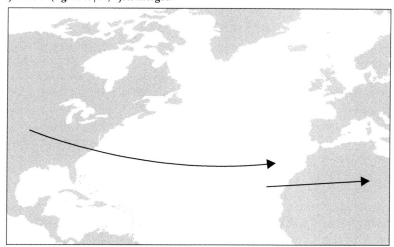

Fig. 12.1. Example jet paths across the North Atlantic in winter.

2009/10, for example, it persisted for much of the winter, spreading snow, ice and disruption all the way from Spain to Finland.

We have, in fact, encountered a situation a little like this before. Twenty-one thousand years ago the jet followed a similar path, in the Last Glacial Maximum when the giant ice cap over North America altered the stationary wave pattern downstream. The ultimate effect of this was to reduce the northward tilt of the jet, and instead angle it straight across the Atlantic. The shifting of the Atlantic jet southward and back again, as the ice advanced and then declined, likely contributed to the raft of climate changes that drove human migrations around North Africa, the Middle East and the Mediterranean during this period.[105]

At the opposite end of the scale, the North Atlantic jet can shift much further north than usual, accentuating the northeast tilt even more. This path carries relatively warm air right up into the Nordic Seas, the zone between Greenland and Norway on the very edge of the Arctic. Some particularly interesting examples of this are the winters of 1815–16, which are years of great historical interest.

William Scoresby was the son of a successful whaling captain based out of Whitby in northern England. He first accompanied his father on an expedition at the age of eleven, but then returned to school and would study at Edinburgh University before taking over command of his father's ship in 1811. In each of the following summers he sailed north, primarily in search of whales but also to quench his thirst for exploration. Thanks to his academic training, Scoresby kept detailed records of the weather conditions during his voyages, which plied the prime whale-hunting grounds along the edge of the sea ice pack to the east of Greenland.

Scoresby had good evidence behind him, then, when he claimed in 1817 that change was afoot in the Arctic. His records from the summer of 1816 showed a sudden warming of the climate compared to the previous years, accompanied by a northward retreat of the ice edge. On encountering similar conditions the following summer, he was moved to write to Sir Joseph Banks, President of the Royal Society: 'I observed on my last voyage (1817) about 2000 square leagues, (18,000 square miles) of the surface of the Greenland seas, included between the parallels of 74° and 80°, perfectly void of ice, all of which disappeared within the last two years.'[106]

Scoresby's report was eagerly received by those in the British establishment who were set on exploration of the Arctic, in particular John Barrow of the Admiralty. With Barrow's keen support, two separate expeditions were launched to the Arctic in the following year of 1818. One, led by David Buchan, followed Scoresby's route east of Greenland in an attempt to reach the pole. Unfortunately, however, they found the climate as cold, and the ice as abundant, as ever. Storm-damaged, they limped home having barely made it past Spitzbergen. John Ross led the other expedition west of Greenland in search of the fabled Northwest Passage, the potential sea route from the Atlantic to the Pacific around the northern edge of Canada, but with no more success than Buchan.

Two hundred years on, and rapid warming really is in progress across the Arctic this time. While still not a regular prospect for commercial shipping, the melting ice has brought the Northwest Passage closer within reach, with even a few cruise ships having successfully made it through. Scoresby was not mistaken though, the dramatic warming he experienced in 1816–17 is quite consistent with the vagaries of North Atlantic climate; in particular, a poleward-shifted wintertime jet stream with a strong northeast tilt. This directs both mild air and powerful mid-latitude cyclones up to the very edge of the sea ice in the Greenland Sea. The resulting loss of winter sea ice paves the way for a (relatively) mild and ice-free summer such as the ones that so surprised Scoresby.[107]

Although it would be many years even until James Glaisher got his first hint of the jet stream, the British could have known something of this kind of variability. The polar expeditions had fuelled the public interest in all things Arctic, and as a result several relevant books were translated into English for the first time. One of these was a Danish text with the catchy title: 'Greenland: being extracts from a journal kept in that country in the years 1770 to 1778 by Hans Egede Saabye, formerly ordained minister in the districts of Claushavn and Christianshaab; now Minister of Udbye, in the Bishoprie of Fühnen; and grandson of the celebrated Hans Egede'.

Saabye's diary of his life as a missionary on the west coast of Greenland provided rich entertainment for Victorian England. As might be expected, there was plenty of adventure, with expeditions to reach neighbouring settlements beset by many hazards and extremes of cold and stormy weather. There are tales of boat trips caught out by storms and perilous crossings of icy and rocky slopes. One particularly harsh winter is recounted in detail, during which Saabye observed the desperate and starved Greenlanders stripping hides off the walls of their homes so they could chew on them for nourishment.

Equally stirring are his accounts of the scandals of Greenland life, from tales of witchcraft to disputes settled by murder and revenge. As a missionary he appears, at least, to have behaved with relative discretion and delicacy on encountering the local tradition of polygamy. He recounted with fascination the customary courtship ritual in which a suitor simply chose his desired bride and then physically dragged her out of her home by the hair. Her compliance in the proposed marriage was judged on a subtle assessment of how violently she protested her fate over the following few days.

With all this excitement, it is perhaps to be forgiven that the Admiralty overlooked a minor, and more sober, detail in Saabye's diary: 'Every winter, in Greenland, is severe; but they are not all equally so. The Danes have observed, that, if the winter in Denmark has been severe, that in Greenland was, in its kind, more mild, and vice versa.'

This casual statement, seemingly reflecting common knowledge in Saabye's community at the time, stands as the earliest written record of what is now recognized as the second most important pattern of natural climate variability in the whole Northern Hemisphere, after El Niño of course.[108]

By the late nineteenth century, serious scientific attention had been paid to understanding how European climate varied, largely motivated by extreme winters such as that of 1879/80, which was immortalized in Monet's paintings of the frozen River Seine. The relationship described in Saabye's diary was soon confirmed by direct temperature measurements. This is now generally known as the North Atlantic 'see-saw' in wintertime temperatures. Quite simply, when it is unusually cold in Greenland and easternmost Canada, it is often unusually warm in northwest Europe; not just in Denmark but from France all the way north to Arctic Norway.

A fundamental advance was made by the Austrian scientist Felix Exner, a central figure in the 'Austrian School' of meteorology, which was probably the most serious competitor to the Bergen School of Bjerknes and colleagues at the time. Exner was in fact the first person to make a successful numerical weather 'forecast'. The fact that this forecast took longer to calculate than the change in the weather it predicted in no way devalues this great achievement, the forerunner of all scientific weather predictions today. During a visit to Washington D.C. in 1904–05, Exner took a snapshot of data from across the United States and, using only the equations of fluid motion, managed to predict on paper how the weather would change over the following four hours, with reasonable success.[109]

Of most relevance to our story is a statistical analysis that Exner published in 1913, which examined pressure and temperature data from across the Northern Hemisphere. This revealed a striking relationship between the pressure measured over the Arctic, and that at lower latitudes, in particular the Mediterranean: when one of these regions experienced high pressure, the other generally had low pressure. Furthermore, Exner noted that the wind changes associated with his pattern of surface pressures could explain the temperature see-saw, as some regions would receive a warmer airflow, and other regions a colder one.[110]

However, it was Gilbert 'Boomerang' Walker who had the honour at least of naming this pattern of variability. Following Exner, he included the whole Northern Hemisphere in his correlation analysis of surface pressure data; the study which had identified the El Niño-linked Southern Oscillation for the first time. Across the North Atlantic he found strongly opposing variations in pressure, which he described as a swaying of atmospheric mass north and south over the ocean. He identified Iceland and the Azores, in the subtropical mid-Atlantic, as the centres of his pattern, which he christened the North Atlantic Oscillation. This name, often shortened to NAO, along with the 'centres of action' of Iceland and the Azores, persist to this day.[111]

Although well established, the word 'oscillation' in this context has often caused confusion. Walker used this term to evoke a swaying of mass from one centre of action to the other. The pressure, as measured by a barometer at the Earth's surface, indicates the mass of air in the atmospheric column above that point. So if the pressure increases and your barometer rises, that means you have slightly more air piled up above you than before. In the case of the NAO, when the pressure rises over Iceland then it often falls over the Azores, reflecting an imperceptible shift of air from south to north over the Atlantic.

However, 'oscillation' has often been misinterpreted to mean that the NAO is periodic in time, i.e. that it has preferred cycles so that it repeats itself at regular intervals like a pendulum. This is not the case; the NAO varies on many different time scales, from week-to-week up to decade-to-decade and beyond, and there is no preferred cycle.

But what really is the NAO, and how does it relate to Saabye's temperature see-saw? The NAO at its simplest is just a measurement: subtract the pressure at Iceland from the pressure at the Azores and that gives you a number which describes something about the atmosphere over the Atlantic. The pressure is normally higher at the Azores than Iceland, so that when this NAO *index* increases then the normal pressure difference is enhanced. Usually, the index is adjusted to be zero on average; then a positive value indicates that the pressure difference is stronger than usual, and a negative value that it is weaker than usual.

Recall that according to geostrophic balance, wind doesn't blow along the line from high to low pressure, as might be expected, but at right angles to this. The pressure difference between two points is therefore related to the wind passing between the points. Hence, the Iceland-Azores pressure difference measures the strength of the wind blowing from west to east across the North Atlantic: when the index is positive, this wind is stronger than usual. Since it only depends on the pressure at the two island endpoints, it is in some way representative of the average wind at all the latitudes between them.[112]

The NAO, then, is really just a simple, ground-based measurement of what the North Atlantic eddy-driven jet is doing. The only small complication is that a given change in the NAO doesn't always mean that the jet is doing the same thing. If the jet is strengthened, say, then the wind between the two islands will be stronger and so the NAO index will increase. So far, so simple. But it turns out that also when the jet shifts northward a little, then the wind falls more squarely between the islands and the NAO index increases again. In this way, the NAO reflects changes in both the strength and the position of the jet. The position of the jet turns out to be more important in determining what the NAO index is, but its strength does also play a role.[113]

As an example of its variability, the NAO was strongly positive in the winter of 2011/12 because the jet was shifted northward for much of the winter. When the jet shifts to a different latitude, it is accompanied by the storm track; the collection of mid-latitude cyclones which ply the North Atlantic. With the storms moved north, Spain and Portugal then endured severe losses in the agricultural and hydropower sectors as a result of the worst winter drought in decades. Even the southeast of England suffered from very dry conditions, as most of the storms passed over Iceland further north, only just clipping the northwest corner of Scotland. After a few relatively dry years, southern England was facing serious risk of drought itself, though this was to be avoided in true British style with one of the wettest summers on record.

The NAO was also positive in the winter of 2013/14, when the enhanced stationary wave over North America was freezing the eastern half of the continent while at the same time keeping California firmly in drought. Downstream of all this, the jet surged dead straight across the Atlantic in a fairly normal location but with truly exceptional strength. The strong, straight jet served as a conveyor belt for storms which battered Europe, leading to flooding in many countries and providing the memorable image of the train tracks left hanging over the pounding waves on England's southwest main line.

The NAO is a very powerful, simple measure of the atmosphere, which reliably emerges as one of the strongest correlations in analyses such as Exner's and Walker's. However, these two quite different examples show that it just gives a broad overview of the atmospheric circulation above the Atlantic, and can reflect a wide variety of flow patterns.

The previously mentioned cold European winter of 2009/10, when the eddy-driven jet shifted far to the south and merged with the subtropical jet, is an extreme example of the negative sign of the NAO. Other classic examples are 1962/63, when western Europe got so cold that sea ice began to form on the coast of Kent in southeast England, and 1941/42, when extreme cold helped to thwart the German invasion of Russia.

These examples show that many of the impacts of NAO variability can be linked directly to the changing path of the jet and the storm track. These control which areas feel the mild influence of the warming southwesterly winds in winter, and also which areas get the storms. But some impacts are less obvious; why, in particular, is there a temperature see-saw between Greenland and Scandinavia?

To understand this, we have to consider the full pressure pattern of the NAO. The two centres of action of the NAO, Iceland and the Azores, each lie roughly at the centre of semi-permanent circulation structures, as shown in Fig. 12.2. The Azores lie close to the centre of the subtropical high that we met the first time Grantley crossed the Atlantic, so much so that this is often called the Azores High. Unsurprisingly, Iceland is near the middle of the Icelandic Low, an average cyclonic flow around the northern North Atlantic. As the jet varies, the pressure changes over the two centres reflect broader scale changes in the strength of these two circulations.

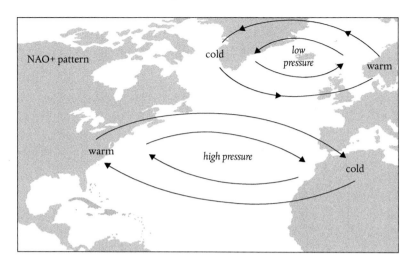

Fig. 12.2. Airflows of the NAO pattern affect temperatures around the North Atlantic basin.

Take the positive NAO, for example, with strengthened westerly winds across the Atlantic. These enhanced westerlies lie at the border between a strengthened Icelandic Low to the north, and a stronger Azores High to the south. Hence, the flow *around* the islands is enhanced, as well as the wind going across the Atlantic.

Importantly, this means that the NAO is accompanied by changes in the north-south winds, and these are very efficient at changing local temperatures by moving air masses across the latitudes. In the positive NAO case, the cyclonic, or anticlockwise, flow around Iceland is strengthened, leading to stronger southerly winds to the east of Iceland and northerly winds to its west. Hence, Scandinavia feels an unusually warm flow of air from further south, while Greenland instead gets a frigid airflow from the north, and this is the root of Saabye's temperature see-saw.

Similar effects are seen further south; as the Azores high strengthens, the enhanced anticyclonic winds around it bring warm air north over the United States, warming the country all the way from the Rockies to the east coast. Across the pond, the winds instead take cold air southward, cooling much of North Africa and the Middle East.

Now we can see why the NAO emerges as the most important pattern of Northern Hemisphere climate variability after El Niño. The variations of the North Atlantic jet are not just felt locally, but all around the North Atlantic basin due to the associated circulations around the Azores High and Icelandic Low. For the full effect, we couple to this the highly variable nature of the jet here: as a purely eddy-driven jet it does not have the constraining influence of the steady Hadley cell, which pins the subtropical component of the jet in place across the Pacific. Variations of the Atlantic jet, as measured by the NAO index, affect temperatures and rainfall across much of the Northern Hemisphere land mass, from the Sahara to the Arctic and from Mexico to Siberia.

By changing the pattern of surface winds and temperatures, the jet variability also has a very important effect on the ocean beneath it, with both the horizontal gyre and the vertical overturning circulations affected. Changes in the temperature and wind-induced roughness of the ocean affect the growth of both phyto- and zoo-plankton, and hence the whole of the food chain above them, to the extent that an east-west see-saw, like that in temperature, has also been observed in the strength of cod stocks. Similar effects are seen in both land and freshwater ecosystems around the basin.

Effects of the jet fluctuations on human life are similarly profound. The associated variations in both rainfall and temperature have impacted agricultural economies around much of the hemisphere. They have contributed to desertification in sub-Saharan Africa and to water supply shortages which have escalated tensions in the Middle East. Other sectors of the economy have also fallen prey to the whim of the jet. As just one example, when the jet dramatically shifted south in 1996, Norway was deprived of precipitation to the extent that it had to buy in coal-generated power from Denmark to compensate for the loss in hydropower.[114]

We are now in a position to learn a little more about Scoresby's experience, which triggered the attempted British push into the Arctic. As we look further back in time we have fewer reliable records of the weather conditions. But based on the observations that we do have, and also climate *proxies* such as records from tree rings and ice cores, it seems likely that the NAO was unusually positive in the winters around 1816. Hence, the Nordic Seas region between Greenland and Scandinavia was likely warmed by a southerly airflow and at the same time ravaged by storm activity. These effects in combination would have hindered the wintertime advance of the sea ice, and likely contributed to the mild and relatively ice-free conditions that Scoresby met in the following summers.

Further evidence for this is provided by the records of the Hudson's Bay Company, comprising ship logs from many voyages between England and the Hudson Bay trading posts.[115] These records suggest that on the other side of Greenland, the same summers of 1816 and 1817 ranked as some of the coldest on record and, in quite the opposite of Scoresby's experience, ships there were regularly trapped by ice on their normal routes. Saabye's temperature see-saw strikes again.

Hence, several lines of evidence suggest that the Arctic ice was not miraculously melting away in the early 1800s, as the British had hoped, but was instead shifting around under the influence of the NAO, a manifestation of the powerful variability of the Atlantic jet. But why, we should then ask, did the jet shift so far north in these two subsequent winters? Is this the kind of thing that can just happen for no other reason than the chaotic and turbulent nature of the eddies that drive the jet? Or was there some other, hidden agent at play which managed to influence the path of the jet itself? This is the question we turn to next.

Although sustained by his daily walking routine, Joseph remained physically delicate. He felt the weather keenly, and would often complain to his correspondents if the winter was particularly harsh. He adopted a strict diet to preserve his constitution, almost entirely vegetarian with abundant consumption of lemon juice. He didn't care for wine and would fuel his long, hard work evenings with hot, sweet tea instead. While his equations remained the only true love of his life, Joseph ultimately began to want for a wife, simply for household purposes of course, to care for him and to make his day-to-day life simpler and more tranquil.

CHAPTER 13

Drivers

At 7pm on 10 April, 1815, the entire mountain of Tambora on the Indonesian island of Sumbawa exploded into flame and ash. The blast was heard clearly even hundreds of miles away. Forests and villages near to the volcano were obliterated within hours, some from falling ash and steaming hot avalanches of volcanic debris, others by ferocious winds generated by the sudden heat. An estimated ninety thousand were to lose their lives in Indonesia alone, making it clearly the deadliest volcano in recorded history. There has not been a larger known volcanic eruption in at least the last 700 years.

Up to 50 km³ of molten rock is thought to have been released into the atmosphere as ash and pumice. Great rafts of pumice were to be found floating in the Indian Ocean thousands of kilometres away. Within a day, the ash cloud in the atmosphere above the volcano had grown to roughly the size of Australia.

Crucially, some of this material made it right up into the stratosphere, the stable zone of the atmosphere extending up for many kilometres above Grantley's level. As much as sixty megatons of sulphur dioxide was injected into the stratosphere at altitudes of around 20 km. Much of this was converted into sulphate *aerosol*, broken up into such tiny particles that it could remain suspended in the air for a year or so before falling back to the ground.[116]

Indeed, the following winter saw bizarre red and yellow snowfalls covering the slopes of Alpine mountains on the other side of the globe. Brilliant red sunsets were remarked upon the world over, as the light from the setting Sun reflected off the underneath of the stratospheric aerosol layer.

Most importantly for the climate, sunlight during the day was reflected and scattered off the top of the aerosol layer, reducing the sunlight reaching the surface by about 1%. As a result, the Earth's surface cooled

149

globally by around half a degree Celsius over the following couple of years. 1816 would go down in history as the 'year without summer', when crop failures and famines spread chaos around the world, from Europe to China, India and North America. The result was a major global food crisis and not one epidemic, but three: typhoid in Europe, the plague in the Mediterranean and cholera in South Asia.

The bleakness of the weather at this time has been credited as a key inspiration for Mary Shelley's Frankenstein, started while she was holed up in a retreat by Lake Geneva, telling ghost stories with her husband-to-be Percy and the celebrity poet Lord Byron. It has also been suggested to have left a lasting impression on the young Charles Dickens, who later displayed a penchant for cold, snowy winters in his novels.[117]

And yet despite the global cooling, and the associated chaos, these were the very years that Scoresby recorded as remarkably *warm* in the Nordic Seas. If, as seems likely, the Atlantic jet did shift poleward to part the sea ice ahead of Scoresby, it must have been a strong enough effect to overcome the volcanic cooling. But could the volcano itself have had an effect on the jet, pushing it so far north in those winters? Can the jet respond to driving by some distant event such as an eruption?

Actually, we have already seen something similar in that the jet over the Pacific can be strongly affected by El Niño. In that case, we can somewhat loosely refer to El Niño as a remote *driver* of variability in the jet. For the case of the volcano, however, the question is significantly harder. To find an answer to this we will have to learn more about the eerily still world of the stratosphere.

Once again, the humble balloon has proved instrumental in unlocking the secrets of the atmosphere. The typical cooling of the air by around 7°C per kilometre of altitude has been known since 1787, when Horace Benedict de Saussure nobly lugged a thermometer and barometer up Mont Blanc to the very roof of Europe. Scientists were hence understandably cautious when they discovered a breakdown in this rule as weather balloons reached higher and higher into the atmosphere.

The German Richard Assmann and Frenchman Teisserenc de Bort could easily have been bitter rivals, but instead became close friends as they carefully improved their balloons and compared data. After many years they convinced themselves there was no fundamental error in their results, which generally agreed despite differences in method and equipment. In 1902, they simultaneously announced their findings to the world: once their balloons reached heights of around 10 km, the temperature stopped falling. (Interestingly, this is around the height that James Glaisher made it up to in his balloon, so perhaps if he had remained conscious for just a few more moments he might have made a truly incredible discovery. . . .)

Now we know that this is the region of the atmosphere where the ozone layer is to be found. Since ozone strongly absorbs solar radiation, especially the ultra-violet part, the temperature actually increases as you ascend through the stratosphere. This explains the incredible stability of the atmosphere here, as the warmer, lighter air is on top. We also know that the temperature varies considerably with latitude in the stratosphere, as well as altitude. In particular, the stratosphere over the winter pole becomes incredibly cold due to the total lack of sunlight, resulting in a much stronger temperature contrast than in the troposphere below, which is warmed from underneath by the Earth's surface.

We therefore have a very strong temperature contrast between the tropical and the polar stratosphere, and in line with thermal wind balance this means that the stratosphere has its very own jet stream. Often termed the *polar night jet*, this encircles the winter pole at heights many kilometres above Grantley's jet, and differs from his in its steadiness: consistent with the stability of the stratosphere this jet generally just goes round and round the pole with minimal variation, see Fig. 13.1.

But not always. Another great balloon discovery was made in Berlin in 1952 by Richard Scherhag. Another scientist who dedicated his life to painstakingly improving his methods and eliminating errors, Scherhag

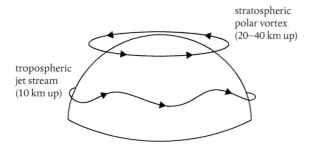

Fig. 13.1. The tropospheric jet stream and the stratospheric polar vortex (also called the polar night jet).

pushed the level of reliable balloon data up to 40 km and maintained careful and systematic observations over Berlin for many years. On 27 January 1952, Scherhag was shocked to discover what he called an 'explosive' jump in the temperature far above Berlin, which rose by tens of degrees in a matter of days.

Originally named 'Berlin events', such occurrences were soon recognized to be truly massive in scale and are now known as *sudden stratospheric warmings*. The north-south temperature contrast in the stratosphere is concentrated about the polar night jet, which encloses a *polar vortex* of extremely cold air. Under normal conditions, Berlin lies at the edge of this vortex, with very cold air high above it.

The explosive jump in temperature during a sudden warming is caused by a major disruption to the polar vortex. The troposphere below is teeming with Rossby wave activity, and the very largest of these waves can travel upwards to disturb the normally steady polar vortex. Sudden warmings occur when the vortex is so disrupted that it is either shifted away from the pole or in some cases completely broken up into smaller vortices (see Fig. 13.2). These events lead to some of the most dramatic and beautiful flow patterns in the atmosphere, and also to profound changes which it takes several weeks for the atmosphere to recover from. The warming itself occurs as relatively warm air from lower latitudes surges north to take the place of the vortex, and the entire polar night jet is temporarily destroyed.[118]

The great importance of sudden warmings now emerges. Having been bashed around and broken up by the troposphere below, the tide now turns and the stratosphere begins to influence the layers below it. The Rossby waves are still moving upwards into the stratosphere, but each one makes it up a little less high than the previous one. Their weakening effect on the vortex hence ratchets down and down through the stratosphere. Once this disturbance reaches the tropopause, at the boundary between the two layers, the weather patterns below are eventually impacted. Scientists still debate the relative importance of different specific processes here, but the interaction between the jet and Rossby waves of different flavours seems key.[119]

The ultimate effect is a weakening and southward shift of the eddy-driven jet stream, a classic negative NAO pattern with the largest signal around the Atlantic basin but impacts spreading to some extent around the hemisphere. Opposite effects are seen when the stratospheric polar

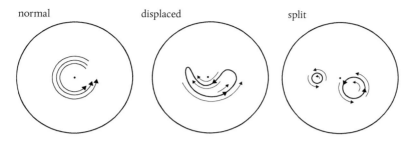

normal   displaced   split

Fig. 13.2. Typical polar vortex disruption during sudden stratospheric warmings, as seen looking down on the North Pole.

vortex strengthens, with a tendency in the following weeks for a stronger and more northward-shifted Atlantic jet. So, even though the stratosphere weighs only a tenth of the weather layer below, its influence can reach down to steer the mighty jet stream. These remarkable events then really do appear to be an important example of the tail wagging the dog.

The whole process of vortex disturbance and downward influence can take a month or so, highlighting a valuable potential source of predictability which can give early warning of all-important jet shifts weeks ahead of time. And if something else affects the likelihood of stratospheric sudden warmings, we may get an even earlier indication. This is where the volcanic aerosol cloud comes into play.

Although shading the Earth below, the aerosol warms the air locally in the stratosphere by absorbing some of the incoming radiation from the Sun. Inside the polar vortex, however, the Sun has vanished below the horizon and so there is no such heating. Hence, the aerosol further increases the temperature contrast from tropics to pole and strengthens the vortex even more. If strong enough, this influence can spread downwards to give a nudge to the Atlantic jet.

As well as volcanoes, other factors might be able to affect Grantley's jet stream through this stratospheric pathway. Some of these are radiative, such as the small variations in the power of the Sun on an eleven year cycle. These changes are negligible in terms of the total solar heating and so are a much smaller influence on global temperatures than the rise of greenhouse gases, for example. However, the variability in ultra violet (UV) radiation in particular can lead to temperature changes in the stratosphere and hence a potential effect on the jet.

Other drivers work by changing the Rossby wave activity which disturbs the stratosphere. El Niño, for example, sends anomalous waves around the hemisphere, so it may be that any influence of the tropical Pacific on Europe doesn't travel there in a straight line but via the polar vortex high above the Arctic.

Another culprit is something called the *Quasi-Biennial Oscillation* (QBO), a magical pattern of winds tens of kilometres above the equator that for once actually does oscillate predictably like a pendulum: east, west, east, west and so on. These winds can steer the path of waves propagating up into the stratosphere, and hence affect the vortex again.

(The terrific eruption of Krakatoa in 1883 makes an interesting diversion at this point. The track of the volcanic aerosol cloud was pieced together from a multitude of eyewitness reports, revealing the presence of strong high-altitude winds encircling the globe in around fifteen days. Some have claimed this to mark the discovery of the jet stream itself, but a closer look reveals an even more interesting story. The cloud travelled west around the world, not east as Grantley has, and stuck largely to the tropical latitudes near the equator, so it was not in the jet stream at all. The volcanic particles had in fact made it up into the stratosphere and signal that a strong easterly phase of the QBO prevailed at the time. Had the great eruption occurred just one year later, the cloud would likely have gone east rather than west.[120])

Before we get too carried away with the power of the stratosphere, we should broaden our range of influences. Yet more potential drivers of jet variability exist that do not need to send their messages to the jet via the stratosphere. Ocean temperatures around the world can vary from year to year and these changes can trigger Rossby waves which are quite capable of making their way to the North Atlantic without taking the high road.

Also, more local temperature changes can influence the jet directly, by altering the temperature contrast which ultimately fuels the storm track eddies. The pattern of ocean temperatures in the North Atlantic itself, right under the jet, are the clearest examples of this, but there may also be effects from the shifting of the sea ice boundary just to the north, or the extent of snow cover over the continents.[121]

We should pause at this point to issue an important health warning. We are entering territory where few things are really certain. The careful reader may have spotted an increased use of words such as 'probably', 'likely', and 'may'. While good scientific evidence has been presented for these mechanisms, there are serious limitations to what can be concluded. A fundamental problem is the lack of good observations as we go back in time, so that only a few decades of sound data are often available to test theories on.

Even when linkages are known with relatively high confidence, the drivers are not guaranteed to deliver the goods every time. Recall the 2015-16 El Niño, which confounded expectations and failed to bust the droughts in both California and Cape Town. Any potential driver only affects the *likelihood* of a change in the jet, it doesn't make it a certainty.

As a result, it is very hard to determine with any confidence what drivers actually had an important effect on the jet in any given year. For cases such as Tambora in the relatively distant past, the challenge is even harder. There is certainly evidence from both observations and computer models that really strong volcanos can nudge the Atlantic jet northward. This clearly could explain the unusual warmth that Scoresby found, even in the global year without summer.

Given the pattern of warming to the east of Greenland and cooling to the west, an NAO-like change of the jet in those winters seems likely. But it is still possible that this occurrence in the years just after the volcano was a coincidence. Certainly not every volcano is followed by a shift of the jet. For example, it seems the eruption of Pinatubo in 1991 did not even affect the stratospheric polar vortex significantly (although Pinatubo was much weaker than Tambora).

Somewhat frustratingly then, the answer to the question of whether Tambora could have affected the jet has to be: yes, it *could* have, but we don't really know if it did. Probably all we will be able to say on this is that given a large number of Tambora-like eruptions, on average the jet would shift north in the following winter.

For more recent events, however, we are starting to be able to make statements with a little more confidence. This is due in particular to the much better observational data in recent decades which provides us with a wealth of information on how the atmosphere and oceans evolved during each event.

The southward-shifted jet which led to the cold European winter of 2009/10 was clearly associated with the influence of a sudden stratospheric warming reaching down to the ground. The warming itself was likely favoured by an El Niño and the easterly phase of the QBO.[122] Other drivers could have played a role, perhaps the pronounced minimum in the solar cycle that year.

The situation was similar in some ways, but different in others, in 1963 when the English seashore began to freeze. No El Niño this time, but Rossby waves from the tropical Pacific likely played a role in any case. Again, the QBO was blowing from the east and a stratospheric warming came at the end of January to prolong the cold.[123]

In contrast, the strong, straight jet of 2013/14, which brought storm havoc to Europe, likely had a helping hand from the tropical Atlantic and an unusually strong polar vortex above. The vortex itself could well have been enhanced by the QBO, this time blowing from the west.[124] Think you can spot a pattern? Well, in the following year (2014/15) the QBO came from the east but, defying expectations, the NAO was positive and the jet strong and shifted north.

Not all cases involve the stratosphere either. The 2009/10 winter was so unusual that it left an imprint of altered ocean currents and temperatures that would linger just under the surface of the Atlantic for the whole of 2010. It then seems that this pattern of temperatures rose again to the sea surface to give a southward nudge back to the jet stream in December 2010. As a result, the Central England Temperature record showed the second coldest December since 1659, and several airports across western Europe were closed at times.[125]

While most meteorological excitement occurs in winter, summer climate is also important and here the behaviour is not dissimilar, with the exception that several of the wintertime drivers are absent. The summertime stratosphere is a very quiet place indeed, with no polar vortex and few disturbances of any kind. But purely tropospheric mechanisms can still be active.

As an example, many of the summers between 2007 and 2012 brought heavy rain and flooding across the British Isles and into mainland Europe. The summer jet normally lies a little north of the winter one, but in these summers it shifted south to target western and central Europe with repeated storms. In a quirk of the summertime circulation, the Mediterranean region to the south was unusually dry, even though the storm track had shifted south towards it. The shifted jet in this string of peculiar summers was likely favoured by the state of the North Atlantic Ocean below the jet, which at the time was unusually warm compared to the oceans elsewhere.[126]

In all of these cases, both for winter and summer seasons, it is likely that *internal variability* also played a role. This is scientific shorthand for 'the jet just changed a bit on its own, with no other secret agent having directed it'. The random effects of chaotic storm activity alone can certainly lead to quite large variations in the jet. As discussed earlier for the case of El Niño, the driver just loads the dice in favour of a certain change. The biggest, most extreme and persistent events invariably occur when the random effect of internal variability pushes in the same direction as one, or maybe even several, external drivers.

The behaviour of the jet in most seasons, in fact, arises from some mixture of internal, random variations and a forced component. This forcing can potentially come from factors such as ocean variations, El Niño and the QBO, which are separate from the jet although clearly part of the overall climate system, or from truly external factors such as volcanos, solar or man-made changes. Crucially, this component of forced variability opens the door to potentially useful seasonal forecasts of the Atlantic jet.

For many years the North Atlantic has been the downfall of seasonal forecasting. In many regions these forecasts are often quite successful, due to the power of El Niño in particular, so we say that the prediction systems are *skilful*. Over the North Atlantic and Europe, however, there is often a conspicuous lack of skill. A breakthrough occurred in 2014, though, when Adam Scaife and his team at the UK Met Office analysed test runs of the latest model and discovered a remarkable level of skill in predicting the winter NAO.

These were encouraging, but tantalizing results. Prediction models are tested by attempting to re-forecast previous years, in a set of what are termed *hindcasts*. In this case, the model was started using data from the beginning of recent Novembers, in an attempt to 'predict' the winters that actually followed. A relatively small sample of twenty years was used for these tests, but out of these the model managed to predict the correct sign of the average NAO index for the coming winter period in something like three-quarters of the cases. This might not sound like a very impressive hit rate, but compared to the dismal performance of systems before this, the improvement was marked.

Since then, the test runs have been extended and the same level of skill has been found in other versions of the same model. Skill has also emerged, to a slightly lesser extent, in some other systems as well.

Remarkably, there is even a bit of skill if the Met Office model is run out through a whole year to predict the NAO in the second winter.[127]

The realization of this skill can be attributed to the hard work of many scientists over the years working painstakingly to improve the computer models, to make the representation of known physical processes more accurate. The continued increase in computer power has also helped, by enabling both the atmosphere and ocean to be represented in finer detail than was previously possible.

This improvement mirrors that which has been exhibited by the same models in making weather predictions on shorter time scales. Weather forecasts, despite being easy targets for ridicule, have consistently improved over the years. This improvement has come at the rate of one day per decade, so that a five-day forecast today is as skilful on average as a four-day forecast was ten years ago. The progress in numerical weather and climate prediction is hence one of the truly great success stories of modern science.

The successful winter NAO forecasts have a profound conceptual, as well as practical importance. By successfully predicting the winter average state of the jet from November, they prove that some significant amount of the jet variability must indeed be forced in some way. Left to its own devices, the Atlantic jet would vary chaotically and become unpredictable within a week or two. The success of the predictions must lie in some external forcing, such as solar or volcanic aerosols, or some climate component which varies more slowly, such as the oceans or the QBO. The powerful and seemingly wildly variable Atlantic jet has actually revealed itself to be steered, and quite strongly at that, by predictable drivers from elsewhere.

This story serves to illustrate a very important aspect of climate science: the forecasting problem provides the ultimate test bed for theories and hypotheses. If a computer model, based only on the basic laws of physics, can successfully predict the state of the atmosphere at some future time then it must be doing something right. Each new year and every extreme event, damaging as they may be, provides more valuable cases which can be used to test and ultimately improve the models.

Unfortunately, a cloud hangs over this otherwise very encouraging success story. While the model is capturing a forced, predictable signal in winter jet variability, this signal in the model appears to be far too weak.

To make a seasonal forecast, the computer model is not just used once to make a prediction of the future, it is repeated many times. The atmosphere is highly chaotic in nature, as first realized by Edward Lorenz, the pioneer of the energy cycle that we met in Chapter 8. Lorenz himself used the image of a seagull flapping its wings, rather than the butterfly which has entered scientific folklore today, but the meaning is the same.[128] The tiniest change in one place in the atmosphere will amplify over time, due to complicated nonlinear terms in the fluid equations, to the point where the whole global pattern of weather can be different.

So for seasonal, and in fact most weather forecasts, a whole collection or *ensemble* of simulations is performed, each with small changes made to the initial state of the atmosphere (the equivalents of the flapping of wings). These different simulations spread out over time and give an indication of the range of possible evolutions of the weather patterns.

Forecasts are hence inherently made up of probabilities: they don't tell us what the weather will do next, they give us an estimate of the risk of particular patterns developing, and also an indication of the overall level of uncertainty. When all the different simulations predict broadly the same pattern, we can often have high confidence in this forecast. These days, this is often the case for country-scale weather forecasting up to about five days ahead.

In other cases the ensemble can be highly spread, with different runs predicting very different futures (see Fig. 13.3). This is the case for the winter NAO forecasts. An ensemble of twenty to forty separate forecasts is typically used to predict each winter, and these generally suggest a range of wildly different winters, with jet behaviour of all flavours and strengths represented. However, if this broad cloud of possible futures is averaged together, the result turns out to be a very weak, but remarkably accurate, shadow of the winter to come.

The model is obviously capturing some predictable signal in the jet, otherwise it would not ultimately get the correct answer as it does. But the balance within the model is wrong: the predictable signal is too weak and the chaotic noise is too strong.[129]

Although this suggests that the forecasts can be made more accurate by simply adding more simulations to the pack, and running the computers more, this issue has raised concerns for the future. These models are the same as, or closely related, to those used to predict the future impacts of man-made climate change. That the models can correctly capture a

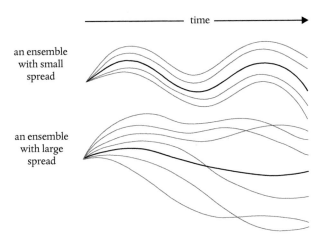

Fig. 13.3. Two ensemble forecasts diverging at different rates, with the average of all the separate simulations shown in bold.

forced influence on the jet is clearly very good news. By applying the climate models to the seasonal forecasting problem we can rigorously test their ability to predict the behaviour of the real world.

But if the response of the jet to external forcing in these models is too weak, then our best forecasts of any jet change under greenhouse warming may prove to be well off the mark.

Ten years after his breakthrough into the world of calculus, Joseph's reputation was well established. Great work kept emerging from his lonely late-night frenzies, such as an elegant new theory to explain the waves that travel along a plucked string. Even the desperately complex behaviour of fluids was starting to yield to his onslaught of equations. Yet all was not well. Joseph was becoming increasingly frustrated that several promised improvements to his position had failed to emerge. Despite all of his achievements, his salary had not increased a penny in all his years of work. Somewhat reluctantly, he began to think seriously about leaving Turin, the city he had lived in all his life.

CHAPTER 14

Exit

London, 1910: streetlamps are twinkling on, puncturing the mist with small spots of light. A blanket of low, puffy clouds gave a gloomy afternoon in the city, which is now deepening further into night. The air is damp but still, and ribbons of smoke rise straight up into the cloud. The tops of the spires are lost in a darkening soup of cloud and smog. But despite the gloom, George Banks is happy and content as he makes his way back to his home in Cherry Tree Lane, after an ordered and productive day in the city.

By the following morning, changes are afoot. Windows of blue sky are emerging through gaps in the cloud, allowing beams of weak winter sunlight to break through. But at the same time, there is a chill in the air. There is a large house at the end of the lane that has been painstakingly made up as a ship, complete with rigging and cannon. A glance at its telescope-shaped weathervane confirms the change: the wind has swung around to the east.

For George Banks, however, more profound changes are on their way. In a matter of days he will have left his job at the bank, disgraced. His household will be in chaos and his children, Jane and Michael, will be running wild around London with a chimney sweep and a flying nanny...

An east wind is certainly not the norm here in Northwest Europe, at the downstream end of the great Atlantic storm track. The wind in London comes from the southwest sector on about half of all days, with the remainder spread more or less evenly throughout the remaining three-quarters of the compass points. Life here has been shaped by the prevailing south-westerly winds which bring mild Atlantic air but also a plentiful supply of cyclones.

Over Europe, Grantley is fast approaching the end of the great jet stream spiral. Passing the British Isles he crosses the fearsome North Sea: one of the busiest of the world's shipping regions despite the frequent gales. Twenty years ago the North Sea oil industry was in its heyday, as Europe played its part in mankind's efforts to pump carbon out of the seabed and into the atmosphere. Today, the oil and gas industry here is well in decline while the offshore wind business grows steadily.

In 2017 alone, 560 new offshore wind turbines were installed across a total of seventeen wind farms off the European coast.[130] Sixty percent of the offshore wind power generated around the entire globe that year came from installations in the North Sea. The engineering challenges are daunting, and seemed insurmountable to many. However, the cost has reduced considerably in recent years and the market is now projected to expand steadily. As just one example, 2017 saw the successful launch of the world's first commercial floating wind farm, 'Hywind'. This consists of five giant turbines located 25 km off the Scottish coast, each one towering 175 m above the waves and yet only attached by chains to the seabed, 100 m below.

Offshore wind presents such an attractive proposition because wind speeds are higher and more consistent over the ocean surface than the land, where all the hills, mountains, forests and cities add up to a rough surface which provides drag and hence slows down the wind. Another factor, of course, is the lack of neighbours to complain about the aesthetics of planned wind farms. But Europeans have been exploiting their abundant wind resource for many centuries, with estimates of the order of a hundred thousand windmills in operation at the industry's peak in the eighteenth and nineteenth centuries, working away to power mills, pump water or drive machinery.[131]

Some have even suggested trying to tap directly into the much more powerful winds up at Grantley's cruising altitude. However, the engineering and politics involved in this are much more challenging than for the case of the floating turbines. Thankfully, the eddy-driven flow here extends right down to the water, so that surface-based turbines do indeed feed off the very underbelly of the jet stream. Studies have shown that abundant potential exists here for energy supply well into the foreseeable future before the additional drag on the jet significantly alters weather patterns.[132]

Across the North Sea, the jet hits the 'lovely, crinkly' Norwegian coast. Grantley may well find himself in the vicinity of Bergen, where Jack Bjerknes drew out his classic cyclone blueprints. As many a visitor to Bergen can attest, this is storm country indeed, with some form of precipitation falling two days out of every three. As well as lying at the end of the path of the Atlantic storm track, the shape of the jet here gives a noticeable boost to cyclones which otherwise may well find themselves tired and spent after crossing the ocean.

Ahead of Grantley lies the great Eurasian landmass that he has crossed once already. The drag from the land surface acts to quell the mighty storm track, as does the relative lack of the moisture which invigorates oceanic storms. Hence, the storm track which gives birth to the eddy-driven flow is dying out, and with the Hadley cell several thousand miles to the south there is no longer a mechanism for maintaining the jet. This, then, is the end of the road for the great current which has carried Grantley around the world. We have reached the *jet exit*.

When a great river finally reaches the sea, the flow is released, free from the constraints of its banks and the force of gravity which has pulled it ever downward from its source. Then the water slows and fans out into the open ocean, swirling and mixing with the salt water. The jet exit is similar: this mighty current of air slows and fans out as it does so, breaking up into strands which intertwine with streams of slower and less organized air over the continent.

The fanning out of the winds at the jet exit can have a profound effect on cyclones here, similar to the behaviour in the right-hand jet entrance region over America. Here, the action is in the left-exit region: the poleward side of the wind fan spreading out over Europe. As before, the spreading of the winds up at jet level encourages air to rise up from underneath.[133] This deep ascent stretches the column of air all the way down to the surface and hence makes it spin faster, just like in our over-used ice skater analogy. If a cyclone happens to drift into the left exit region, it suddenly gets a helping hand from above which stretches it upwards and re-invigorates its circulation (see Fig. 14.1).

Several examples of this behaviour occur every year. One particularly notable case was that of storm Kyrill which barrelled across Europe on 18 January 2007. Over the preceding two days the storm had crossed to the north side of the jet over the mid-Atlantic, then by a fluke of alignment it tracked the left-exit region of a 230 mph jet streak extending straight towards Europe. By the time it struck the continent, the central pressure had deepened to below 970 mb, and had done so rapidly enough to qualify as a 'bomb' cyclone by deepening at least 24 mb in a

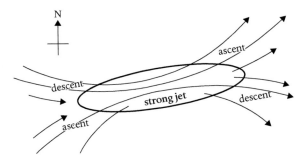

Fig. 14.1. Cyclones are strengthened in the areas of ascent which lie under the right-hand entrance and left-hand exit regions of a jet streak.

twenty-four-hour period. The storm brought hurricane force winds to many parts of Europe, claimed at least forty-six lives and uprooted an incredible sixty-two million trees, mostly in the low mountain ranges of central Germany.[134]

Another well-known example is cyclone Lothar, which remained weak as it crossed the Atlantic rapidly over Christmas 1999, then exploded in the left exit region just as it hit Europe on Boxing Day.[135] The storm blasted a trail of devastation across France, Germany, Switzerland and Austria, leaving over fifty dead and triggering insurance claims totalling eight billion US dollars.

Severe European cyclones such as these rank amongst the most costly of natural disasters for the insurance industry, topped only by large earthquakes or hurricanes in regions such as America or Japan. Both Kyrill and Lothar were especially damaging as they came not alone, but in clusters of several strong storms striking in a short period of time. Kyrill was just one of six named cyclones to hit Europe during the period of 6–20 January of January 2007, with total insured losses of $10 billion. Lothar was preceded by Anatol earlier in the month and then followed, on the very next day, by Martin, with a combined cost from the three of $18 billion and over 150 lives.

The clustering of cyclones in this way is another example of the controlling hand of the jet stream. In both of these cases (and other similar ones) the jet was strong, steady and stuck in position, guiding several subsequent storms to the same weather-beaten area.[136]

Frequent and recurrent storms are therefore a fact of life in Europe, at the end of the jet stream. In addition, the jet also has one final trick up its sleeve. What really happens when the wind swings around to come from the east; how can it blow directly against the mighty jet that has carried Grantley around the world? In fact, not only can the east wind hold its own against the jet, it can do so for a remarkably long time. In both the book and film, Mary Poppins, the flying nanny, arrives on the east wind and stays until the west wind returns, with plenty of time for adventures in between.

The author Joseph Conrad wrote about the battle of the winds in his memoir of his sailing days, *The Mirror of the Sea*:

> There are no North and South Winds of any account upon this Earth. The North and South Winds are but small princes in the dynasties that make peace and war upon the sea. They never assert themselves upon a vast stage. They depend upon local causes—the configuration of coasts, the shapes of straits, the accidents of bold promontories round which they play their little part. In the polity of winds, as amongst the tribes of the earth, the real struggle lies between East and West.[137]

Conrad pays homage to the west wind as ruler of the seas surrounding Europe, albeit with a note of respect to the storms it brings. The east wind, however,

> an interloper in the dominions of Westerly weather, is an impassive-faced tyrant with a sharp poniard held behind his back for a treacherous stab... I have seen him, like a wizened robber sheik of the sea, hold up large caravans of ships to the number of three hundred or more at the very gates of the English Channel.
>
> (Ibid.)

Conrad was plagued by bitter memories of one instance during which he claims to have been stranded for six whole weeks in the English Channel at the mercy of the east wind, with his ship's crew starving near to death for want of just a couple of days of west wind to blow them home.

East winds seem to have been feared for centuries, as evidenced by an old English proverb 'when the wind is in the east, 'tis good for neither man or beast'. The French writer Voltaire, exiled in London for 1726–2, declared

> This east wind, is responsible for numerous cases of suicide . . . black melancholy spreads over the whole nation. Even the animals suffer from it and have a dejected air. Men who are strong enough to preserve their health in this accursed wind at least lose their good humour. Everyone wears a grim expression and is inclined to make desperate decisions. It was literally in an east wind that Charles I was beheaded and James II deposed.[138]

Extended periods of easterly winds in the mid-latitudes are synonymous with a process known as *blocking*, which we mentioned briefly in Chapter 5. During these events, the jet stream itself and the usual train of cyclones which ride along it are literally blocked by a large and immobile weather system. The cyclones, and the jet stream winds themselves, still charge across the Atlantic towards Europe, but on reaching the block they are deftly diverted to the north or south.

Blocks often, but not always, feature a giant anticyclone, revolving majestically in the very path of the jet. We have conceptual pictures of blocks, just like the cyclone blueprints that Bjerknes drew up, but these are a little more varied. While meteorologists would almost always agree on whether a given feature qualifies as a cyclone, there is still often some disagreement over whether a particular event should be considered a block.

While blocking events therefore come in several different flavours, one type dominates in Europe at least (Fig. 14.2). And by most metrics that scientists have developed, Europe emerges as the region of most frequent events. Amongst the mid-latitudes at least, Europe is 'Blocking Central'.

A typical European block develops as follows. At first, out over the Atlantic, the jet will likely be strong and fostering vigorous storm activity. These storms grow as Rossby waves, undulations of the jet north and south. So the jet snakes its way towards Europe, and as it gets closer the waves get larger, swinging further to the north and then back even further to the south. What occurs next truly counts as high drama on the jet stream: blocking is what happens when the Rossby waves break.

Flash back, briefly, to the waves crashing onto the beach at Soup Bowl. As they enter the bay, the waves roll gently up and down. Closer in to

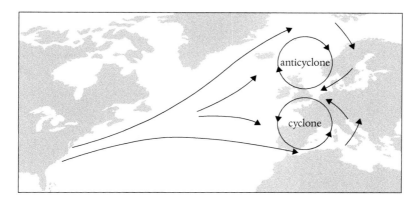

Fig. 14.2. A typical blocking structure at the end of the jet over Europe.

shore, the peaks rise higher and the troughs fall lower. Then, in the last few seconds, the very summits of the waves start to overturn, forming graceful overhangs of water which tower above the gliding surfers.

Breaking Rossby waves do exactly the same thing: they overturn. Except in this case it doesn't happen in the vertical direction, but the horizontal. In the place of the wave peak, we have a northward meander of the jet. So overturning in this case means that this whole meander stretches north and then leans over to the side, entrapping to its south a mass of air which just now was sat happily on the polar side of the jet (see Fig. 14.3).

Before long, the jet has gone from being gently wavy to overturned. Follow the line of flow along the very core of the jet and it will now trace out a gigantic backwards 'S' shape in the air (ƨ); detouring to the north and then to the south. Crucially, in the middle, the flow turns back on itself. For a brief moment of its existence, it's like the jet itself is blowing from east to west.

In this classic picture of European blocking, there is indeed an anticyclone; it lies in the northern part of the S, with an equally important cyclone to the south. Sometimes, the overturning goes the other way, i.e. like a normal 'S', though this is rarer in Europe. Either way, in the middle of the block, against all the odds, there is a region of easterly wind that can strand fleets of ships for weeks and seems to have been very dangerous for English kings.

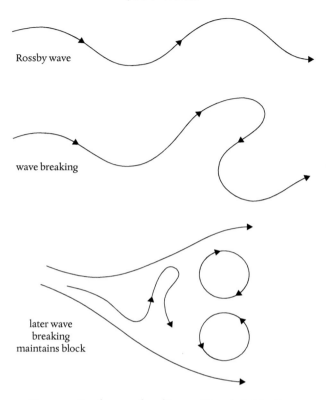

Fig. 14.3. Rossby wave breaking and its role in blocking.

One of the defining features of blocking is persistence, as Conrad discovered to his peril. Once overturned, the jet stream is effectively broken, and takes a while to reassert itself. Blocking events typically last a week or two, so most often shorter than Conrad's epic delay, but a few very long lived events have wreaked havoc over the centuries.

Far from breaking it down, new storms approaching from across the Atlantic often actually reinforce the blocking structure. At ground level the weather can seem static, with a broad anticyclone stuck fast and easterly winds set in. But up at jet level the flow is highly turbulent, with each new wave stretching out as it approaches the region and breaking in turn, repeating the process that started the block in the first place.

Two scientists, Noboru Nakamura and Clare Huang from Rossby's old department in Chicago have memorably described blocking as a traffic jam of waves on the jet stream. The jet exit over Europe is prone to this, as the weakening and spreading of the jet there reduces its ability to carry waves. This acts like a constriction in the road, and as the waves pile in toward Europe congestion can be inevitable.[139]

But why is blocking and the associated east wind such a bad thing? After all, at least temporarily it acts to shield Europe from the onslaught of Atlantic storms. In many ways, blocking gives calm, settled weather as a result. But it does often lead to extreme weather, mostly from its effects on temperature. We know that north European winters are warmed by the south-westerly prevailing winds associated with the jet. When blocking sets in, this warm air current is diverted away. Instead, the winds are weak or even worse they are reversed, in the infamous easterlies which can bring frigid air from the continental interior.

By a twist of fate, the situation in summer is reversed but equally hazardous. In summer, the westerly wind off the ocean acts as a mild, cooling influence, as the land has heated up more rapidly than the sea. So when blocking sets in, this mild influence is removed and temperatures can soar. Changes in cloud cover can also contribute: anticyclonic conditions often come with clearer skies which mean long freezing nights in winter and long scorching days in summer.

Blocking likely played a role in the long, hot summer of 1666 which, as recorded by Samuel Pepys in his famous diary, left London perilously dry. When the 'Great Fire' broke out in a bakery on Pudding Lane, the flames were fanned by a strong wind. The devastation tellingly spread out westward from its source, revealing the hand of the dreaded east wind.

Blocking, then, leads to extremes in both seasons. For many locations in Europe, blocking will often be to blame for both the coldest week of winter and the hottest week of summer. Blocking over Europe led to the coldest weeks of the fearsome winters of 1963 and 2010, and also the deadliest periods of the heatwave summers of 1976, 2003, and, of course, 2010 in Russia.

During the writing of this book, in fact, blocking in late winter of 2018 brought cold winds which were memorably branded 'the Beast from the East' by the British media. Yet just a few months later, blocking struck again, this time giving the hottest periods of the 2018 European summer heatwave.

By bringing winds from unusual directions, blocking has occasionally had some unexpected impacts worthy of the name 'freak weather'. One example occurred in autumn 2012, when easterly winds associated with a block over the North Atlantic fatefully steered Hurricane Sandy back westward towards New York.[140]

Two years earlier, in 2010, another Atlantic block blew the ash cloud from the Eyjafjallajökull volcano southeast from Iceland, instead of northeast as the usual prevailing winds would. The result was the greatest disruption to air traffic since the Second World War, affecting ten million passengers and grounding planes across Europe and as far south as Morocco.[141]

But how will blocking affect our favourite air passenger? The jet often splits around a blocking pattern, so for Grantley this is an uncertain time; he is approaching a fork in the road. He might end up in the southward fork, pushed down over southern Europe towards the Mediterranean.

Or, if he takes the northern path then he will truly enter new territory. Having headed mostly eastward throughout his circumnavigation of the globe, he could now find himself crossing lines of latitude at quite a pace. The northward branch of the jet rides up around the northern flank of the blocking anticyclone, often reaching the Arctic regions of Scandinavia or Russia, or even beyond.

Having started his journey on a tropical island, and weathered both the subtropical desert belt and the turbulent mid-latitude storm zone, Grantley could find himself heading over the ice and, being a winter balloon, out of the Sun. Here it is drier, and the winds calm as the storm track is left behind, but the cold can be intense. Up at jet level the average winter temperature is around $-60°C$, one of the coldest regions of the whole atmosphere.

In this scenario, Grantley could be part of one final, important story. Let's say that he is diverted north into the Arctic by a blocking system over Europe, but let's also suppose that he is not alone but accompanied by a cyclone; a typical, mid-latitude storm of the type that Bjerknes drew, also diverted north by the block. While the great white expanse of the Arctic would hardly notice a small but distinguished weather balloon floating high above it, the entrance of the cyclone would have quite an impact.

The winter sky in the deep Arctic is often gloriously clear. The colder the air gets, the less moisture it can hold, so there's often little water vapour around to form clouds. We know from experience that clear nights are generally colder than cloudy ones. This is because the cloud absorbs radiation emitted by the surface and re-emits some of it back down again. In the Arctic winter, with little cloud and also little or no sunlight to bring energy in, the temperatures plummet.

The cyclone, however, is a huge swirling mass of much warmer, mid-latitude air. By Arctic standards, this air carries with it a huge amount of water vapour, some of it evaporated from the ocean surface in the warm subtropics far to the south. When the cyclone bursts into the scene, it suddenly floods the Arctic region with moist air. On contact with the much colder native air, the moisture rapidly condenses into droplets, just like on a cold window pane, and soon huge areas of the Arctic can be cloaked in a protective, warming blanket of low cloud. Under such circumstances, the surface temperature can easily rise by over $10°C$.

So the same blocking event that gives some of the coldest days in Europe can also be responsible for some of the mildest days of the Arctic winter. The Arctic has been warming rapidly over the last couple of decades. Man-made climate change is a key underlying cause of this, but the exact mechanisms by which the Arctic in particular has warmed so much already are still under debate. Blocks, cyclones and the guiding hand of the jet stream seem to be playing some role.[142]

Just as Joseph's loyalty to Turin was finally wavering, he received a wonderful invitation; to move to Berlin and take up a position in the highly esteemed Academy of Sciences. In terms of pay, prestige and independence, this offer was a clear improvement on his current position. It was an invitation to join one of the best-regarded academic institutions in Europe, at just the time that he wanted it most. And, despite of all this, he declined. The reason he gave was simple, yet confounding. Leonhard, the mathematician that Joseph most admired in the whole world, was also based in Berlin, and despite all the years exchanging letters brimming with mutual respect, Joseph declared 'it seems to me that Berlin would not be at all suitable for me while he is there'.

CHAPTER 15

Future

Now that we have reached the end of Grantley's whirlwind ride around the Northern Hemisphere, it is time to tackle the elephant in the room: is the jet stream changing and, if so, are we to blame for it?

The scientific consensus on the basics of climate change is clear: the planet is warming rapidly due to man-made greenhouse gases. In the global average, temperatures are already one degree warmer than they were a hundred years ago. The planet is warming much more rapidly than it does in any natural cycles, and this can only be explained by our emissions of carbon dioxide and other gases.[143]

But how is this affecting the jet stream? What does climate change mean for the wind patterns of the world? Is the jet getting weaker and wavier, as some have claimed?

To address these questions, we will first spend this chapter imagining a future world. Many effects of climate change are only just starting to emerge today. So for a clearer picture of how the physics of the jet is changing, it helps to consider a future with much stronger warming and hence a much stronger signal of change.

To do this we lean heavily on our climate models; those computer-rendered versions of the physics equations that grew out of weather forecasting models.[144] But also we rely on simpler models and careful theoretical analyses to help us understand and assess the predictions that the climate models make. Then, once the impacts of a strong warming have been understood, we can return to the present to see if any hint of these changes might be apparent already.

Let us start by thinking into the future, say to the end of this century. We don't know how much warmer the world will be then since, apart from anything else, this depends strongly on our actions now. In the worst case scenario, tellingly referred to as business-as-usual, we could be 4°C warmer than today on average. But if we act now to aggressively limit the emission of greenhouse gases the damage could be much less.[145]

For the sake of argument then, we'll consider some future great-grand-daughter of Grantley, cruising around a 4°C warmer world in 2100. Will we still need weather balloons in eighty years time? Who knows, but the detailed level-by-level measurements they provide have not been beaten yet by even the best satellites.

One of the most obvious changes, in addition to the extra heat of course, will be that the waves will crash noticeably higher up the beach at Soup Bowl than they do today. It's hard to say how much exactly, but a conservative number is that the average sea level might be half a metre higher than it is today.

A significant fraction of sea level rise is simply due to the expansion of sea water as it warms; physics that is known to many from school experiments heating metal bars in the lab. This effect is important because of how well it is understood, providing a useful lower bound on predictions of sea level rise. But other effects such as the melting of polar icecaps will also be very important contributors to sea level rise and are much harder to calculate.

Now let's launch our new, turn of the twenty-second-century balloon. Another clear difference is that, if we want to fix its altitude to cruise at the level of the jet peak, just like its predecessor, then we will have to aim a little bit higher. The jet is expected to extend upwards in a warmer world, so that the level of maximum wind may be a kilometre or so higher than today.

This is caused by changes in how the atmosphere radiates heat. As greenhouse gases increase, more of the radiation emitted by the planet is immediately reabsorbed by the atmosphere just above it, leading to warming of the air. High up in the stratosphere, however, the changes will actually lead to an overall cooling. Here, the extra carbon dioxide effectively makes the stratosphere more efficient at losing energy to space, and this is why the air cools.[146]

The tropopause is the boundary between the troposphere below (where temperature decreases with height) and the stratosphere above (where temperature increases with height). With warming below and cooling above, the net effect is to push this boundary up a little, and the jet maximum moves along with it (Fig. 15.1).[147]

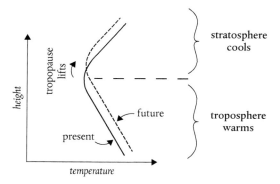

Fig. 15.1. Lifting of the tropopause under climate change, due to warming below and cooling above.

Before long, our new balloon will be racing across the Atlantic as the jet picks up speed, and another key change will have become apparent. The atmosphere does not warm everywhere at the same rate, and one region which warms particularly strongly is the upper part of the troposphere in the tropics. So while everywhere is warmer, the air on the right-hand side of our balloon's track is especially warm.

This is a simple consequence of the moist physics of the tropics. One of the most fundamental effects is that warmer air can hold more water vapour, and so as air warms its moisture content increases. The tropics is the warmest part of the atmosphere, so there is considerably more moisture in tropical air than there is in air elsewhere. As a result, a column of air here behaves differently under warming from a column outside the tropics.

Crucially, the tropics is structured so that the temperature falls with height at a rate given by the theoretical *moist adiabatic lapse rate*. But this rate is itself sensitive to the amount of moisture, in that the temperature falls off less rapidly with height if the air is moister. If we start at the surface and move upwards the air will still get colder as we go up, just not quite so quickly. The upshot is that the warming effect of climate change gets amplified as we ascend in the tropics. Our warming of 4°C at the surface will be more like 8°C up at jet level (Fig. 15.2).

The steady rise in moisture as the air warms in fact gives rise to some of the clearest global patterns of climate change, such as in the

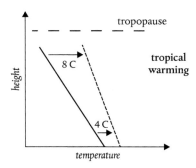

Fig. 15.2. A column of air in the tropics will warm more strongly at upper levels than lower down.

distribution of rainfall. In terms of moisture, what the Hadley cell does is systematically transport water towards the equator. This is because there is much more water in the warm air near the surface than in the colder air aloft, and this warm, low-altitude air is carried equatorwards by the cell.

So the Hadley cell today moves water from the subtropics into the deep tropics, giving rise to the desert belt around twenty to thirty degrees north that we first encountered over Mauritania. Unfortunately, the future here does not look particularly promising. In a warmer world, the Hadley cell becomes more efficient at transporting moisture. The same strength winds will carry more moisture simply because there is more in the air. This results in an increased transport of moisture from the regions which are already dry to those which are already wet. Hence the wet regions get even wetter and, more worryingly, the dry regions get even drier.[148]

These are examples of *thermodynamic* changes, meaning that we don't need to think about how the world's wind patterns will change in order to understand them, just the basic physics of temperature, pressure and humidity in a gas. This means that scientists have relatively high confidence in these changes, because the physics is simple and well-understood. There are still some uncertainties, however, for example the wet-get-wetter effect seems to hold on average but not necessarily everywhere, particularly over land regions.

Another well-known thermodynamic effect is that land areas will warm more strongly than the ocean surface. So if the planet is 4°C warmer on average, the actual damage will be more like 2–3°C over the ocean and 5–6°C over land. This effect is again due to land-ocean differences in lapse rate, i.e. the rate of decline of temperature with height.[149]

In contrast to the thermodynamics, questions which relate to the dynamics of the atmosphere are notably harder.[150] By 'dynamics' here, we mean the winds, forces and accelerations which make up the fluid dynamics equations. These features have such complexity that simple pen and paper theories are few and far between. Possible changes in the circulation systems and wind patterns of the world are therefore much less certain than the thermodynamic changes. Unfortunately for us, the jet stream lies firmly in the dynamical domain.

For predictions of how the jet will change, we therefore have to rely more than we would like on the answers churned out by the supercomputers. One of the clearest dynamical signals to emerge from these is that the whole of the Earth's tropical belt is likely to expand. The Hadley cell will stretch just a little bit further away from the equator than it does at the moment. By the time Grantley Junior makes it to Japan, she'd likely be further north than Tsukuba, though only by a couple of hundred kilometres or so at most.

While subtle, the expansion of the tropics and associated poleward shift of the subtropical jet is one of the most robust predictions of the computer models. Although it seems like a small change, the effect on particular regions could be huge. In Chapter 4 we saw that civilizations in the African Sahel, on the border between the wet tropics and dry subtropics, have been vulnerable to small shifts of the boundary zone in the past. Similarly, present day communities along the edges of these climate zones face potentially devastating changes over the coming century if the models are right.

Once past Japan our balloon will hit the storm track as before. How will things here be different? Again, the increase in moisture will be an important factor. Many of the most damaging storms to hit the midlatitudes have been known to have been invigorated by moisture. As air rises up within a cyclone, the moisture condenses out into rain and this process heats the air, making it even more buoyant and so speeding its upward motion. Does this mean, then, that with more moisture in the air, the storms will be stronger? The answer, somewhat surprisingly, seems to be no.[151]

To the atmosphere, moisture is a form of energy. By taking water vapour evaporated from the ocean in the warm lower latitudes and transporting it polewards before it condenses out, the atmosphere has effectively moved some heat poleward. (Energy is used to evaporate water at low latitudes and then ends up heating the atmosphere when the condensation happens at high latitudes.) So, if two identical storms were to occur, one today and one in 2100, the latter would transport more energy poleward simply because it carries more moisture.

But this poleward energy transport is precisely the reason the storms are there in the first place, so the change has essentially made the storms more efficient in achieving their job. Hence, we expect less storm activity in the future, not more, as fewer and/or weaker storms are needed to achieve the same overall poleward heat transport.[152]

The storm tracks of the future will therefore likely be slightly quieter places in some way, either with fewer storms, or weaker storms, or a mix of the two. It is often said rather loosely that there will be more extreme weather in the future, but mid-latitude storms may well turn out to be a counterexample, with fewer incidents of severe winds in particular. At the same time, however, these storms will probably give greater rainfall, following the wet-get-wetter rule. Also, it is only in the hemisphere average that models often predict weaker storm activity, not necessarily everywhere. As we shall see soon, some regions may actually see an increase in storms, despite the global trend.

For the jet stream as well as the storm tracks, the crucial ingredient in the climate change problem seems to be this underlying need for poleward energy transport. This, after all, is the single most fundamental reason why the jet and the storms are there to start with: to balance the energy budget, the atmosphere needs to move energy from the tropics towards the poles.

We have just said that in the tropics, up at peak jet level, the warming will be stronger than elsewhere. Surely this means that the all-important contrast between tropical and polar temperatures will increase; does this not imply stronger jets and storms in the future, after all?

We now arrive at one of the central causes of uncertainty over how the jet will change. In the upper troposphere, the warming is indeed strongest in the tropics, as we have just heard. Down at the surface, however, the situation is precisely the opposite.[153]

Another clear and very robust feature of the climate model predictions is that surface warming will be strongly amplified in the Arctic, not the tropics. This arises because of local feedbacks, for example the loss of sea ice which uncovers the relatively warm ocean beneath.[154] Winter sea ice coverage is expected to decline by around 30% by the end of the century, and in summer this number is closer to 90%. Some completely ice free summers are expected by this point in the strong warming scenarios, a very dramatic change indeed since the days of William Scoresby.

The result, then, is a veritable battle for the future of the jet stream (see Fig. 15.3). While the upper level changes driven by the tropics act to strengthen the temperature contrast, the lower changes driven by the Arctic act instead to weaken it. The future changes to the jet stream predicted by climate models are actually quite small, but this is at least partly due to a stalemate in the tug-of-war for the jet between tropics and pole. If our predictions of either of these two regions prove wrong, the changes in the jet may be much stronger than expected.

It is not, however, a complete stalemate. One of the two effects does seem to be stronger than the other. And the winner, to the best of our current knowledge, is the tropics. Many of the predicted changes in the atmospheric circulation can be traced back to the amplified warming high up in the air over the equator.

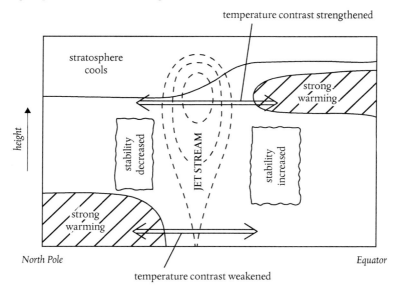

Fig. 15.3. The regions of strong warming in the atmosphere can affect the jet by altering both horizontal temperature contrasts and vertical stability.

In line with the increased equator-pole temperature contrast driven by the tropics, we can therefore expect a slightly stronger jet in the future, particularly in the regions dominated by subtropical driving. Junior will likely complete her journey a bit faster than Grantley did. But the clearest effect is actually a shift of the jet rather than a strengthening. By warming the tropics, we will push the zone of sharpest temperature contrast a little bit closer to the pole, and the jet will move in tandem with this.

Crucially, the amplified upper level warming spreads a little bit away from the deep tropics, even into the dry subtropical desert belt. The precise mechanisms by which this actually affects the jet and storm track are still not completely clear. Different theories involve detailed changes in features such as the size, shape, speed or trajectory of individual storm systems.[155] As discussed in Chapter 9, the jet and storm track are tightly coupled together; an example of fluid dynamical 'eddy-mean flow interaction'. Does the change in the jet cause the change in the storms or vice versa? We would like a simple chain of causality, such as 'A causes B, causes C'. However, when B also influences A, and C affects both A and B, then such simplicity is the stuff of dreams.

Broadly speaking, however, it seems that changes in the subtropical region will be key. Today, this region is brushed by the southernmost limbs of mid-latitude storms, as they reach down to lower latitudes to extract momentum from the Hadley cell. In the future, however, this region will become a less favourable environment for these storms, as here the distribution of warming makes vertical air columns more stable and horizontal temperature contrasts weaker. The storm tracks hence contract a little away from the tropics, concentrating their energy a little closer towards the pole. In a way, it is not really a tug-of-war for the jet after all, more a push-of-war, since whichever region warms the most acts to push the jet away from it.

So the storm tracks contract poleward, and the jet moves along with them. This can also help to explain the expansion of the Hadley cell. The extent of the cell is determined not just by the tropics, but also by the rate at which the storm tracks can pick up the energy and momentum from the cell and carry them away. As the storm tracks move poleward, the edge of the Hadley cell hence has to follow suit.

These are the broadest, planetary-scale changes in wind patterns that we should expect by 2100. Once averaged around the hemisphere, all models predict a poleward contraction of the mid-latitude circulation.

So on average, at least, regions currently located to the north of the jet and storm track may well become stormier in the future, while places to the south may find themselves deeper in the subtropical dry zone. But the uncertainty here really is considerable. The average shift is of a degree or two of latitude, but there is a large spread between the actual numbers that come out of different models.

The situation is even worse if, for some selfish reason, we are not satisfied with the hemispheric average, and want to know how much the jet will shift at our particular longitude. For many regions, there is no consensus between the models over how the jet will shift at all. In some places, the jet is even predicted to shift the other way, due to some regional effect. How the jet will change in your neighbourhood is down to a complicated mix of region and season, but in all cases the prediction comes with a large errorbar.[156]

One particular region which does not seem to fit the overall pattern is Europe, particularly in winter. Here, many models don't predict a particular shift of the jet and storm track, but instead suggest that these may extend slightly deeper into the continent than they do today (Fig. 15.4). The potential for strengthening and extending of the storm track deeper into Europe has serious implications for storm and flood risk in several countries. On the flip side, however, this does lead to a reduction in the occurrence of high-impact blocking events, at least in the models.

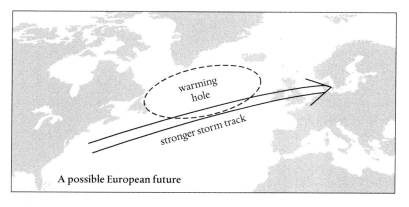

Fig. 15.4. How the changing ocean circulation may affect European climate.

There is some evidence that this distinct local response might again be partly due to the tropical warming. However, in addition to the global battle between tropics and Arctic, there are other, complicating factors in this region of the world. One is the potential for change in the stratospheric polar vortex, whose seasonal wobbles tend to affect this region most strongly. Another potential wildcard is the ocean circulation.

As discussed in Chapter 11, it is a misnomer that the Gulf Stream may shut down under warming. But the models in general do predict a weakening of the overturning circulation; the Hadley cell-like system which carries warm water northward in the Atlantic Ocean. A Hollywood-style catastrophic collapse seems unlikely, but even a moderate weakening of this system leads to a noticeable 'warming hole' in the northern North Atlantic, as the greenhouse-induced warming is locally offset by a reduction in the warm water flowing in from further south. The result is a local strengthening of the temperature contrast on the southern side of the warming hole, and this is one of the local factors driving the stronger storm track into Europe.[157]

On top of the problem that models don't agree, we are also faced with a dilemma over how much faith to put in them. The models are based on sound physics but they are only approximations, limited by our knowhow and the available computer power. They are clearly not perfect, particularly for dynamical changes such as the jet shift. Maybe the models are all missing some subtle change in the moisture fuelling of mid-latitude storms, for example. Or perhaps some change in the El Niño system which sends Rossby waves around the world. The changes described in this chapter represent our best informed guesses of the future, but we certainly can't rule out some potential surprises emerging.

While the models have steadily improved over the years, many of them still suffer from common problems, or *biases*. For the jet, this is normally manifested as an overly 'zonal' circulation. This means that, while in reality the jet stream took Grantley on a graceful spiral around the hemisphere, in the models it is too much of a ring directed straight from west to east, and is often too strong as well. Relatedly, models often underestimate the occurrence of the high-impact blocking events that bring the east winds. If the models are in error over the future occurrence of blocking, this could be very serious.

From a purely practical viewpoint, biases such as this pose a particular problem. Scientists are increasingly being pressed for more detailed predictions of how the climate will change in particular places. A classic example concerns hydrological impacts, for which we would ideally like to know the changes in rainfall likely to be seen in each individual river catchment. But the climate projections are based on models which, in the case of the worst biases, can have the jet pointed at the south of France rather than Scotland. Producing trustworthy local predictions is clearly a great challenge in cases such as this.

Ultimately, how much we trust such models comes down to how much we believe the physics they simulate, and to what extent we can test them against the real world. In this regard, the close link between the climate models and those used for weather and seasonal prediction becomes invaluable. We can learn a lot by testing models in a forecasting framework, where their predictions of the future can be rigorously scored against the hard truth of what actually happens. Not all processes on the centennial time scale of climate change can be assessed in this way, but some usefully can. We finish this chapter with a few words of wisdom from Jack Bjerknes, the pioneer of both cyclone and El Niño research:

> But yet I would give the highest recommendation to the less narrow and more basic field of meteorology, which was the concern of the founders of our science, and which still is our first duty to society: weather forecasting. All too frequently, students, and professors too, shy away from the subject of weather forecasting and go into one of the nice little research specialities which are less nerve racking, and which do not force you to show the public how often you are wrong.[158]

Joseph's future was now bearing down on him with considerable momentum. Although he had perplexingly declined the offer from Berlin, his fate seemed decided. The very next year, Leonhard himself announced his departure from Berlin, to take up a new position in St Petersburg. In his parting exchange with Frederick, King of Prussia, Leonhard was unequivocal. To maintain the high standing of the school of Berlin, and continue the great tradition of mathematical progress that he had started, the king should make all efforts to secure as his replacement the modest Joseph from small-town Turin.

CHAPTER 16

Changes

Changes lie ahead, then, for the great spiral jet stream. Overall, we expect it to shift slightly closer to the pole, but there remains considerable uncertainty over what will happen at any given location. Given this uncertainty and the uncharted world into which we are venturing, can we say anything about the climate today? Has the jet already been affected in some way by the rising tide of greenhouse gases? Several recent extreme weather events have been mentioned in this book; did any of them result from climate change?

First, we start with an easier question: is the planet already warming? The answer is an unequivocal yes. In the instrumental global temperature records, which piece together thermometer measurements from around the world since 1850, the warming is striking. At the time of writing in 2018, seventeen of the eighteen warmest years on record have occurred since the clock ticked over to the new millennium in the year 2000. As Bjerknes urged, we should not shy away from the subject of making actual forecasts. In this regard, climate models have been used to predict global temperatures for several decades now, and the warming predicted by the early models has indeed come to pass.[159]

However, what matters for the jet is not the overall strength of warming, but its pattern, especially the battle between the two warming hotspots of the lower-level Arctic and the upper-level tropics. These are features that have been predicted by the climate models; but have they actually been observed in reality?

For the case of the Arctic, the answer is a firm yes. The polar region as a whole has been warming at least twice as fast as the global average over the last forty years, with the sea ice cover shrinking by about 4% over each decade. Amplified Arctic warming, as predicted by the models, is clearly already here.

Head south to the tropics and the answer is again yes, although this has not been clear until recently. It has been a major challenge to compile trustworthy, long-term records of the air temperature high up above the surface. Measuring upper air temperatures is of course one of the main aims of Grantley and his fellow radiosondes. But this network was designed to detect developing weather changes, not to provide trustworthy long records of climate.

To measure if the world is warming, we need to *homogenize* the records, i.e. account for any changes over time in how the measurements have been made. For example, the records from the weather station at Grantley Adams International Airport show that the model of radiosonde used was changed between 1986 and 1988, and then also the computer system changed in 1991. Accounting for changes such as these is painstaking work, and even then there are likely many other changes which went unrecorded.

Satellite data, while offering many benefits, can suffer from similar drawbacks. When one satellite retires and is replaced by a new one, there is often little or no overlap time, when both are working at once, during which their measurements can be cross-checked against each other. Yet long records of temperature changes over several decades require precisely this stitching together of readings from the old and the new satellites.

Upper air temperature records constituted a serious climate controversy in the 1990s, as early attempts at long satellite records suggested that the air at altitude had not in fact been warming as much as that at the surface. Two decades later, after much meticulous work and, of course, lots of extra years of data, the controversy has been resolved. Once all of the many uncertainties are taken into account, there does not appear to be a fundamental inconsistency between the warming predicted by the models and that actually observed.[160]

So the world is warming, but the tropics are warming fastest at upper levels and the Arctic is warming fastest at lower levels. The battle for the jet stream is underway. And it seems that the tropics might have taken an early lead.

Warming is expected to lead to a widening of the tropical belt and hence a poleward shift of the subtropical, Hadley-driven parts of the jet. Several different analyses of the observational data suggest that the tropics have indeed been widening on average, at a pace of about 20 km per decade.

When you think about it, this a very gradual change. You could easily drive in an hour the distance that the tropics have expanded in the whole of the last forty years. It is just the very beginning of a longer-term expansion, one that has the potential to displace civilizations in the vulnerable boundary regions between climate zones. But crucially, in answer to the question 'has our behaviour affected the jet stream already?', the answer is very likely yes.

But all is not quite as simple as it seems. Even this gradual expansion is actually proceeding at a rate twice as fast as that predicted by the climate models. The reason for this seems to be that not all of the observed tropical widening is due to man-made climate change. Some of it just happened naturally.

Earth's climate is never static, it is always varying. As Walker and others discovered, there are a myriad of patterns, constantly shifting, growing and decaying in the winds of the atmosphere and the currents of the oceans. Just as El Niño describes dramatic changes in the Pacific from year to year, other patterns reflect subtler changes from decade to decade.

Over the last forty years it happens that the distribution of heat across the vast Pacific Ocean has shifted a little, to give relatively cooler temperatures in the tropical East Pacific and warmer ones elsewhere in the basin. This natural swing has, coincidentally it seems, provided an extra nudge to widen the tropics a little faster than predicted.[161]

This does not detract from the role of greenhouse gases, however. Climate change has likely driven part of the observed tropical widening, just not all of it. In fact, it is only by including the response to greenhouse gases as well as the effects of natural variability that the observations and the climate models can be seen to agree.

But this example does serve to highlight the importance of natural variability. It has the power to make or break a climate change signal in these early years of warming. And as we leave the tropics and enter the mid-latitudes, this power only increases.

El Niño notwithstanding, climate variability is generally weakest in the tropics. Wind patterns are dominated by the steady overturning of the Hadley and Walker cells, as opposed to the chaotic dance of jets, eddies and waves which pervades the mid-latitudes. Surprisingly then, although the surface is warming fastest in the Arctic, the warming signal is actually expected to emerge earliest in the tropics, since that is where the natural background 'noise' is lowest.[162]

As we follow the jet spiral into the mid-latitudes, we therefore enter a much more variable, noisy system. Our ability to say anything meaningful about an emerging climate change signal in the jet stream falls off sharply. Observations over the past century show changes in the jets from decade to decade which would easily outweigh the signal expected from man-made warming. By some model-based estimates, it could still be several decades before any sign of jet change emerges from the background noise in the mid-latitudes.[163]

To give just one example, the wintertime North Atlantic jet was relatively strong in the early decades of the twentieth century but then weakened for a few decades around the middle of the century before recovering again.[164] Such slow variations, which likely reflect some interaction with the ocean currents, comprise a source of noise which could easily be mistaken for an emerging climate change if only the last few decades are considered.

It is clear, then, that if only short records of climate are available, we are on dangerous ground. But in some cases, such as when probing the potential effects of amplified warming in the Arctic, that is all we have; the Arctic change itself has only become apparent over the most recent few decades. As discussed in Chapter 10, longer records and sound support from climate models will be needed before any emerging effect of the Arctic on the jet stream will be clear.

But what of all the extreme weather the jet stream has brought us in recent years; has this all just been an act of nature? Over the North Atlantic at least, there is some evidence that the jet stream has been particularly variable recently, as exemplified by the extreme southward shift in the 2009/10 winter and extreme northward shift in 2011/12. But variability itself seems to come and go, and there were equally variable periods earlier on in the twentieth century.[165] So was climate change actually to blame for any of the extreme events discussed in this book?

Scientists used to issue a blanket response to this question: we can't blame climate change for any individual event. 'Climate' is some carefully defined average state, albeit a changing one, and individual extreme events are 'weather', which is different. But it is becoming increasingly clear that we can, and should, say more than this. The way to do this is to resort to probabilities, and consider how climate change has altered the *risk* of certain extreme events.

There are several good analogies for this. One of these is to consider a loaded dice that has been biased to throw a six more often than it should. Let's say you roll the dice and get a six. Did you get that six because the dice was loaded or would it have come up anyway? Of course, we can't really answer that, but we can say that loading the dice increased the chances of getting that outcome.

Alternatively, imagine a baseball player who starts taking steroids and then scores on average 20% more home runs than he did before. As for the dice, we can't say that any individual run was caused by the steroids, but we can say that the probability of his getting a home run at any particular time has increased by 20%.[166]

A rapidly growing area of climate science is focused on estimating these probabilities. This was first attempted for the European heatwave of 2003; a summer that was likely hotter than any other in the preceding 500 years and which led to thousands of deaths across several countries. A pioneering study led by Peter Stott at the UK Met Office compared simulations of a climate model performed with and without greenhouse gas emissions included. This concluded it was very likely that the probability of Europe experiencing a summer this hot had been at least doubled by climate change.[167]

The 2003 heatwave sent a shockwave through Europe. This was the moment that many there woke up to the dangers of climate change. But then just seven years later this event was blown out of the water by the heatwave of 2010 that devastated western Russia and flooded the Swat river valley downstream.

In terms of statistics, this event was much more unusual than the 2003 heatwave. Even by the end of this century, after several more decades of warming, summers this hot will only occur about once per decade, according to climate models. In contrast, summers like 2003 could easily be the norm by 2050, and in the second half of the century they would even stand out as unusually cold.[168] The reality will of course be strongly affected by how quickly we can reduce our use of fossil fuels.

The 2010 heatwave obviously spurred much interest among climate scientists. Several studies have examined different aspects of the event but one key area of disagreement soon emerged: to what extent was climate change to blame?

Some concluded that climate change played a relatively minor role; the heatwave was clearly caused by the persistent Rossby wave pattern that spawned repeated blocking events, as we saw in Chapter 5. The background warming due to climate change might have added a small amount of extra heat, but what's the difference really between 39°C and 40°C?

Others, however, focused on the probabilities and claimed that the risk of such an event had in fact been very significantly increased because of greenhouse warming. Which of these views is correct? The answer was given by a team led by Friederike Otto from the University of Oxford: surprisingly both are.[169]

The strength of the heatwave was indeed mostly natural, and can be blamed on the winds and the waves along the jet stream. Climate change just added the icing on the cake to make the temperatures a little higher than if the same Rossby wave had occurred in the past. But for rare events like this, a small shift in strength can have a huge effect on the probabilities. An extra degree of heat can shift the probability of an event like this from being really tiny to just small, see Fig. 16.1. In their analysis, the probability of such a heatwave had actually tripled since the 1960s. This type of event used to be expected to occur naturally once every ninety-nine years, but now it should be expected once every thirty-three years.

Just as was the case when we considered future changes in Chapter 15, all of our confidence here lies in the thermodynamic domain. It is what

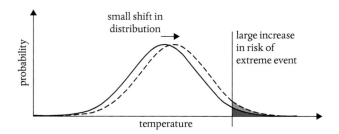

Fig. 16.1. This distribution shifts to the right, indicating a small change in the average value, but a large change in the risk of extreme events.

we call the dynamics (the wave, the anticyclone and the blocking events) which came together to cause the bulk of the heatwave. But the only effect of climate change that we understand was on the thermodynamics: the peak of the heatwave was amplified a bit because the background temperatures are just a little bit higher than they used to be. We could have had the same blocking patterns back in the 1960s, but the air inside the blocks would have simply been a little less hot.

The contrast between dynamics and thermodynamics raises difficult questions concerning current events and climate change. Most of our confidence and understanding of likely changes rests on the thermodynamics, yet the dynamics is crucially important. The risk of an event like the Russian heatwave was significantly increased because of the thermodynamic effects of climate change. But it wouldn't have happened at all if it hadn't been for Rossby's waves along the jet stream.

As a result, there are several different opinions on how we should put numbers on the current effects of climate change.[170] Some simply say that all weather events today have been affected by climate change, as they have developed in what is a new, partly artificial climate.

Others argue that we should just stick to statements about the thermodynamics. The dynamics are simply too hard and uncertain and we shouldn't trust the computer models to get this right. Hence, our statements for any particular event should be similar to those given for the Russian heatwave: if this same weather pattern had occurred in the past, or in an imagined world without human influence, then the thermodynamic impacts of that pattern would have been different in some way.

This latter approach has been that taken most often in practice, apart from anything else because the climate models used in these studies typically do not suggest that many clear dynamical changes have yet taken place. As well as the heatwaves, several other of the extreme events described in this book have been studied in this way.

For example, remember the cold European winter of 2009/10, when the Atlantic jet shifted south and exposed vast areas of the continent to bitterly cold and snowy conditions. The behaviour of the jet was truly exceptional that winter, so that by many metrics there has not been a more extreme season like it on record. In contrast, the temperatures were not so unusual, though clearly cold enough to cause great disruption. By one measure, it was only the thirteenth coldest European winter since 1949.

The reason, again, is the background warming due to man-made climate change. Europe was cold that winter because the usual mild influence of the jet across the Atlantic was replaced by winds bringing colder air from the Arctic, or from the interior of the Eurasian continent. But this air was just not as cold as it would have been if the exact same winds had occurred in the past.[171]

Another example is the long California drought of 2011–17. It is not clear that the rainfall deficit was affected by climate change; the primary cause was of course the shifted jet and storm track. But again, the air was slightly warmer than it would have been in the past, and this exacerbated the drying of the land surface.[172]

All of these examples have centred on temperature changes, which are clearly the easiest to identify in a globally warming world. But the effects of greenhouse gases have been felt in patterns of rainfall as well, with suggestions of effects on individual events. One of the most well-known cases is the flooding in England and Wales in autumn 2000, which damaged ten thousand properties and cost an estimated £1.3 billion in insured losses. In this case, it seems the risk of such an event had been significantly increased by the wet-get-wetter thermodynamical signal of climate change.[173]

Overall then, there is ample evidence that the emerging thermodynamic signal of climate change has already acted to load the climate dice. It significantly increased the risk of several of the most extreme weather events that we have seen in the last few decades.

Increasingly, however, it seems clear that if we really want to put actual numbers on risks and probabilities, we need to try to say something about the dynamics as well. If there was a change in the likelihood of a dynamical event, say a particular weather pattern, this could change the picture completely. Our estimates of the current level of risk of particular events could be very wrong without including this.

Given the level of uncertainty surrounding changes in the dynamics, this is a very daunting task. In many cases the climate models simply do not agree on the response of the dynamics to climate change. Even in the cases when they do agree, we are not sure if we should trust them because they have known deficiencies. However, one recent event in particular opened the door for the first exploratory studies on the role of changing dynamical weather patterns.

Some of the story of the freak winter of 2013/14 has been told in earlier chapters, from the freezing polar bear in the Chicago Zoo to the continued drought in California. Many of these impacts arose from an amplified pattern of stationary Rossby waves. But we also mentioned quite different problems on the other side of the Atlantic, which was battered by repeated mid-latitude storms following an unusually strong, straight jet which extended deep into Europe.

Severe flooding and storm damage occurred in several European countries, and some remarkable rainfall records were established. As just one example, this remains the wettest winter in the whole of the England and Wales Precipitation series, which started back in 1766. The iconic Thames Barrier, designed to protect London from flooding by storm surges coming up the river, was completed in 1982. From then until summer 2017, it has had to be closed 179 times in total due to flood risk, and of these a full fifty occurred in the 2013/14 season alone.

As before, it seems likely that human-induced climate change played a role in this event through the thermodynamic, wet-get-wetter effect. The air within storm systems is warmer than it used to be, so any given storm releases that little bit more water than it would have done if exactly the same storm had occurred in the past. But in addition it is possible that, as some studies have claimed, the effect of human influence can be seen not only in the rainfall but in the jet itself.

Certainly, if we ignore the rest of the globe and focus only on the European region, the behaviour of the jet and storm track in that winter does indeed resemble our best estimate of the local response to climate change. Out over the Atlantic, and indeed in many other regions, climate models show little agreement over how the jet will change under the tug-of-war of competing influences. Over Europe, however, the models show some consistency in predicting a strengthening and extension of the jet deeper into the continent, as discussed in Chapter 15.

The standard approach to test the role of human influence is to compare two sets of climate model simulations, with and without the greenhouse gases that we have added to the atmosphere over the decades. Such an experiment for 2013/14 was performed by Nathalie Schaller and colleagues from several institutions across Europe.[174] The thermodynamic effect was clear, as expected, since the warmer air held more moisture than before.

But in addition, these experiments did indeed suggest that the greenhouse warming so far had increased the likelihood of an unusually strong jet stream. The team performed thousands of simulations of the winter, and on average the jet extended deeper into Europe in the simulations including greenhouse gases than the simulations without. The difference was small, much weaker than what the jet actually did in reality. But we could not hope for climate change to explain all of the event, just the straw that broke the camel's back: the extra little bit that makes it just slightly worse in terms of magnitude, but on the other hand significantly more likely in terms of risk.

So it is just possible that, despite all the uncertainties, we have already witnessed a human-induced change in the jet stream contributing to extreme weather. But the uncertainty is indeed very high.

Just because several climate models agree on this particular jet stream response to climate change does not mean that they are right. The models are in many ways very similar to each other; many of them share the same problems and all use similar assumptions to strip the immense complexity of the climate system down to something that can be simulated on a computer. The computer version is infinitely simpler than the real thing, however big the computer.

Much more work is required before we can be confident that climate change will affect the jet in this way, and more again before we can be sure that these changes have started already. For a start, these early studies will need support from others using different methods and different and better computer models.

Crucially, we will also need a deeper understanding of the exact mechanisms leading to this response. As mentioned in Chapter 15, there is evidence of several different ways that the extension of the wintertime jet and storm track into Europe can occur. It could be due to changes in the tropics, such as the broad, high-altitude warming or the more local changes that can trigger far-reaching Rossby waves. Or it could be the strengthening of the north-south temperature contrast under the jet as the Atlantic Ocean currents adjust. There may even be other, as yet undiscovered mechanisms at play.

There are many reasons to be cautious. But at least for Europe, the wet and stormy winter of 2013/14 might just have given us an early taste of the jet stream behaviour to come.

On 6 November, 1766, at just thirty years of age, Joseph was formally installed as Leonhard's successor as Director of Mathematics at the prestigious Berlin Academy of Sciences. He was warmly and enthusiastically welcomed by most. His talents were eventually recognized even by the few who initially resented the appointment, as their superior, of someone young enough to be their son.

Joseph enjoyed many years of highly productive and uniform life in Berlin, and indeed in Paris after that. He maintained his strict daily routine of afternoon walks and solitary evenings of work. His carefully controlled diet and his quiet, ordered household were attentively maintained by both his first and his second wife, in turn. And through everything he continued his great exchange of mathematical letters with Leonhard. Too different from each other to become real friends, the two men nevertheless maintained great respect and admiration for each other. They never met in person.

CHAPTER 17

Confession

We have reached the end of the story of Grantley and the jet stream. On our journey we have seen how variations in the jet have been responsible for many of the most extreme weather events on record. Some of the recent events have had especially severe impacts because of the emerging effects of climate change, adding extra heat to heatwaves and rainfall to floods.

The jet itself will ultimately change as the planet warms; overall we expect it to get a little stronger and to shift slightly closer to the pole. We are just beginning to see some hints of these changes, particularly in the parts of the jet which are most strongly tied to the tropics.

But now it is time to face up to an awkward truth. We are overdue a confession: a key element of the tale we have spun is quite possibly incorrect. We claimed to have launched a balloon into the very start of the jet stream over Barbados, and then ridden the jet as it spirals around the whole of the Northern Hemisphere. The truth of the matter is that if we really did launch such a balloon, it may well not actually take this path. Grantley's journey as we have told it is certainly possible, but it is unlikely.

The problem isn't one of altitude. We have imagined that Grantley stayed stuck at the level of the jet core as he circled the globe. This feat is by all means achievable, given some clever engineering. In fact, some balloons like this were actually used around the 1950s as a novel way to collect valuable measurements of the upper atmosphere. They helped to reveal the complex twists and turns of the high-altitude winds that would inspire Rossby's theory of the planetary waves.

These were *transosondes*, short for trans-ocean sondes: balloons carefully balanced to reach a certain height and then stay there. As they followed the upper winds they could be tracked from the surface by use of the radio signals they sent back.

197

At the height of their popularity, transosondes were used operationally to provide data for American weather forecasts. Between September 1957 and April 1959, the United States Navy launched 230 of these from the naval air station at Iwakuni in southern Japan.[175] These were giant weather balloons, 12 m in diameter and capable of carrying 150 kg of instruments for several days. The balloons would gradually leak helium, and so to maintain a constant height they would begin their journey with another 200 kg of ballast that could be released to compensate. The trajectory of each balloon would be tracked by a network of radio stations across the Pacific and North America, revealing the path of the upper level winds.

The balloons launched in November 1957 proved particularly interesting. 'Flight 44' actually made it from Japan all the way over North America and to the British Isles in just four days, suggesting an average speed of 120 miles per hour. Unfortunately, this and other successes prompted 'complaints from certain nations', and before long the flight duration was curtailed to avoid interference with trans-Atlantic aircraft. Without these restrictions, the balloons could have lasted up to seven days and stood a chance of making it most of the way around the globe.

It is the paths traced out by the transosondes which reveal the problem with Grantley's story. In all cases these started off in the same way, following Ooishi's jet eastward over the Pacific. But before long, the paths began to diverge. Some forged straight across the ocean, while others meandered far to the north and south. The balloons were launched into a jet which was constantly varying; sometimes shifting north, sometimes south, and sometimes snaking around the highs and lows of a giant Rossby wave.

As a result, balloons launched on different days could well take very different paths across the Pacific, depending on the movement of the jet. But some differences are even more extreme than that. Some of the balloons drifted off and left the jet stream current altogether, often disappearing into the tropics and out of radio range. Several slipped out of the jet into the so-called 'transosonde graveyard' off Baja California, never to be seen again. Those that did make it across the American continent took wildly diverging paths in the Atlantic: some passed Greenland, bound for the Arctic, while others drifted south into the subtropics (Fig. 17.1).

So this is the problem with Grantley's story. It is possible that he wouldn't in fact make it all of the way around the jet stream spiral.

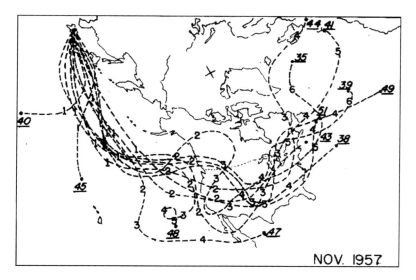

Fig. 17.1. Paths taken by transosondes released from Japan in November 1957.

Instead, it is quite likely that at some point he would part company with the jet.

To understand this, let's consider one final analogy for the jet stream. In a way, the jet is like a great highway. You could picture the iconic American Route 66 or England's M1 linking London in the south with Leeds in the north (or indeed any major highway you are familiar with).

Now imagine yourself sat in a traffic control centre in front of an array of screens. Data floods in from regular sensors all along the road, providing counts of the vehicles at each point. From this, the computer builds up a picture of the average traffic flow, showing which segments of the road are flowing freely and which are moving at reduced speed. The picture you see is of a constant stream of cars following the road, for the sake of argument let's say from London to Leeds. (In this thought experiment you are just following the traffic in one direction; our jet stream is a one-way highway.)

But closer inspection would show something considerably more complicated. The collection of cars that arrive in Leeds at the end of the road are not exactly the same as those that joined it at its start in London. Clearly not every car on the road goes all the way from London to Leeds.

Some of the original set will have left early on, at Milton Keynes for example, while others will have made it all the way to Sheffield, just short of the end. Similarly, many of the cars arriving in Leeds will have started somewhere along the way, perhaps Nottingham or Chesterfield. Still others will have joined and then left again, only taking the M1 from Luton to Leicester for example.

Zooming out on the traffic map would reveal a complex jumble of side roads and junctions. These carry traffic dispersing away from the highway to a multitude of different destinations, but also other flows of new cars converging on to it. In our analogy, these side roads represent other weather systems and flow patterns in the regions alongside the jet.[176] Just as the traffic along the M1 is constantly mixing with that on the side roads, the air does not surge along the jet stream in isolation, but instead mixes and interchanges with the flow around it.

The jet stream is hence not a conveyor belt, constantly transporting air parcels from one end to the other. It is a highway made up of air parcels joining for a while and then leaving to ride the flow of other weather systems and circulations. Some air parcels do follow the whole length of the jet, from Barbados all the way around the spiral to Scandinavia, but these are in the minority.

To study such behaviour, we can use a technique known as *trajectory* analysis, which is applied to a database of wind speed and direction throughout the atmosphere. This involves picking an air parcel and then tracing out the path that it would follow, as it is blown around under the action of the winds recorded in the database.

A fundamental result is that, due to the chaotic behaviour of the atmosphere, the trajectories of two neighbouring air parcels will eventually diverge from each other. The two might track alongside each other for several days, but ultimately they will drift apart and could easily find themselves on opposite sides of the world within a week or two. Hence, we should not expect an air parcel, or a weather balloon for that matter, to reliably follow the whole length of the jet, even if expertly launched right into its centre.

A particularly revealing experiment was performed by Olivia Martius of the University of Bern. She systematically identified air parcels in the subtropical jet and traced out their trajectories backwards, rather than forwards, in time, to see where they had been in the seven days beforehand. Overall, she found that only half of the air parcels had remained in, or relatively near, the jet over the preceding week, during which time they had made it most of the way around the globe.[177]

This might suggest a 50/50 chance for Grantley to complete his journey, but given that he also has to make it out of the subtropics that Martius focused on, and ride the turbulent storm track across the North Atlantic, the odds are almost certainly lower than this. We can now see what happened to those transosondes with the unusual trajectories: they simply took an exit road at some point and left the jet stream. In the same way, a real-life Grantley may well turn off early before reaching the end of his circumglobal highway.

What other possible Grantley stories might there have been then? He could have been caught in an anticyclonic block such as those that caused so much trouble over Russia in 2010, trapped for days to spin round and round in a side-eddy off the jet stream. Or perhaps he could have been bumped out of the jet by turbulence as the mighty Himalayas scraped the belly of the great highway in the sky. Or maybe pulled south by some unexpected airflow, possibly even across the equator and into a whole other hemisphere of wind systems that we have hardly mentioned.

More likely than not, his exit would have been a dramatic one; swept out of the jet by one of the powerful storms that ravage the Pacific and Atlantic Oceans in winter. Hurled aside by the tempest, he could find himself wandering the side roads of the atmosphere for a while.

Or perhaps he just took a slightly different path, for example slipping around the south side of the Rockies rather than the north, before regaining his route. He could even have taken an early shortcut: perhaps plucked from the subtropical Atlantic by the trailing limb of a mid-latitude storm before he even passes the Nouakchott fish market, instead to be pulled directly up to the finish line over Scandinavia.

There are clearly unlimited possibilities for the trajectory that Grantley might actually have taken. What we have really done in this book is not follow the path of a particular balloon, but instead follow the average route of the jet stream. If we take one of our datasets of global winds and average these over several years to get the typical wind speed and direction at any location, the result maps out the smooth, clear spiral that we have followed around the world.

The picture of the jet we have actually followed is therefore effectively like the traffic recorders monitored from the control centre. Rooted to the spot, these just see the flow of traffic going past them, just like if we fix ourselves at a particular location and consider the average winds there. In meteorology, and fluid dynamics in general, this approach of fixing ourselves in space and watching the fluid flow past us is termed an *Eulerian* approach.

If we actually want to do the Grantley experiment for real, to launch a balloon into the flow and see where it goes, we are instead following a different description of fluid dynamics known as *Lagrangian*. In this approach we formulate the equations as if we are moving with the flow, riding with an air parcel just as if we were following a particular car, no matter what turns it takes.

These two approaches are named after Euler and Lagrange, two pioneering eighteenth century mathematicians whose equations of fluid dynamics laid the very foundations of modern meteorology (and much else besides). Both approaches are sound mathematical descriptions of how Newton's laws control the flow of fluids, and both are incredibly useful, they just lend themselves to different problems.

Let's say we wanted to understand the average conditions at a specific location, perhaps the temperature in a given city, the wind over a particular mountain or ocean currents past a certain headland. In this instance we would be fixing ourselves in place, watching the world go by. This would be taking the approach conceived by Leonhard Euler.

But if instead we wanted to launch an imaginary weather balloon and follow it wherever the winds of the world will take it, then we would take a different approach. We would follow in the footsteps of Euler's successor and long-term correspondent, the man he never met in person – one Joseph-Louis Lagrange.[178]

GLOSSARY

Aerosol Tiny particles of a solid or liquid suspended in the air

Angular momentum The appropriate form of momentum for circular motion, which importantly is affected by the radius of the motion

Anticyclone High-pressure weather system which spins clockwise in the Northern Hemisphere

Blocking A persistent weather pattern which blocks the jet and storm track

Buoyancy The upwards force acting on an object in a fluid, such as a boat in water or a balloon in the air

Climate model Computer representation of the equations which govern climate, which can be used to simulate past, present and future climate

Convection The process of fluid rising as it is more buoyant, such as in a pan of water heated from below

Coriolis effect The apparent deflection of motion due to the rotation of the Earth

Cyclone Low-pressure weather system which spins anti-clockwise in the Northern Hemisphere

Dynamics The physics of forces and movement

Easterly Wind blowing from the East towards the West

Eddy A fluid dynamical swirl; weather systems can also be viewed as eddies

Eddy-driven jet A jet which gets its momentum from the transient weather systems

Ekman Scientist who used the three-way balance between pressure, Coriolis and surface friction to explain many features of the climate system

ENSO (El Niño-Southern Oscillation) Atmosphere and ocean disturbance in the tropical Pacific that affects weather around the globe

Extratropics The region of the globe outside of the tropical band

Geostrophic balance The dominant balance of horizontal forces, in which the pressure and Coriolis forces largely cancel each other out

Gulf Stream A fast, narrow current of warm sea water off the east coast of North America

Hadley cell Overturning circulation in the tropics, with air rising near the equator and moving towards the poles

ITCZ Inter-Tropical Convergence Zone; band of ascending air and high rainfall along or near the equator

Jet stream A strong, narrow, current of wind

Kinetic energy Energy that an object or fluid has because it is moving

Latitude Measures how far north or south you are

Longitude Measures how far east or west you are

Mid-latitude cyclones Large cyclones that get their energy from the temperature contrast between latitudes

Momentum Mass times velocity; objects have more momentum if they are heavier or move faster

North Atlantic Oscillation (NAO) Pattern of variability associated with variations of the Atlantic eddy-driven jet

Parcel A small volume of fluid used as a concept in fluid dynamics, e.g. by considering the different forces acting on an air parcel

Potential energy (gravitational) Energy that an object or fluid has because it is at high elevation, so the energy can be released if it falls

Pressure force Air feels a force pushing it from higher pressure to lower pressure, like water along a pipe

Radiosonde A small package of instruments carried up by a balloon to record atmospheric data

Rossby wave A wave of alternating cyclonic and anticyclonic weather systems

Storm track The collective body of mid-latitude cyclones which follow the jet stream

Stratosphere The stable region of the atmosphere above the troposphere, roughly 15–50 km up

Subtropics The band of latitudes at the edge of the tropics, around 20–30° latitude

Subtropical highs Large, near-permanent anticyclonic weather systems covering the subtropical oceans

Subtropical jet A jet stream driven by the poleward movement of air at upper levels in the Hadley cell

Thermal wind balance A large-scale balance; stand with warmer air to your right and the westerly wind will be strengthening upwards in the air above you (in the Northern Hemisphere)

Thermodynamics The physics of temperatures, pressures and densities

Trade winds Steady easterly winds found near the surface over much of the tropics

Tropopause The boundary between the troposphere and stratosphere

Troposphere The active weather layer of the atmosphere, reaching from the surface up to about 15 km

Vertical wind shear How the wind speed and direction is changing as you go up

Vorticity A local measure of spin in fluid dynamics; can be planetary (due to Earth's rotation) or relative (fluid spinning with respect to the Earth)

Westerly Wind blowing from the West towards the East

NOTES

Chapter 1

1. For more on atmospheric observations, see http://www.wmo.int/pages/prog/www/OSY/Gos-components.html, or *A Very Short Introduction to the Atmosphere*, by Paul Palmer (Oxford University Press).
2. For an overview of the science of weather forecasting and its progress over the years, see Bauer, P., Thorpe, A., & Brunet, G. (2015). 'The quiet revolution of numerical weather prediction'. *Nature*, 525(7567), 47–55.
3. Radiosonde data in particular is organized in the Integrated Global Radiosonde Archive (IGRA), and is also included in hybrid model-observation datasets known as reanalyses. See Durre et al (2006): 'Overview of the Integrated Global Radiosonde Archive'. *Journal of Climate*.
4. See *Storm Surge* by Adam Sobel (Harper Collins), for a more in-depth account of tropical cyclone science.
5. Data from the Grantley Adams balloons, and others around the world, is publicly available from the NOAA IGRA: https://www.ncdc.noaa.gov/data-access/weather-balloon/integrated-global-radiosonde-archive.

Chapter 2

6. This is from *Meteorology*, by Aristotle, Book Two.
7. The word Meteorology itself comes from the Greek metéōros, meaning lofty or high, and -logia, the study of. Hence it is the study of things in the sky. Aristotle was not the first of the Greek thinkers to study this, or the last. His student and ultimate successor Theophrastus actually had much better ideas in this area, reverting to the view that wind was simply a moving air current. He felt the damp sea breezes blowing ashore in Athens and was clearly troubled by the idea that this was a dry exhalation from some far-distant land. Unfortunately, it is Aristotle's work which survived from this period, rather than Theophrastus'. See *The Riddle of the Winds* by W. S. Kals for more discussion on this.
8. Mediterranean climate, particularly in winter, is characterized by many local wind patterns caused by land-sea contrasts, and frequent synoptic storms. For an overview see Lionello, P. (2012). *The Climate of the Mediterranean Region: from the Past to the Future*. Elsevier.

9. *Science and Civilisation in China*, by Joseph Needham, Volume Three.

10. Translations of Columbus' letters and logs (which he wrote in the third person) are readily available, for example at https://archive.org/. For a longer account, see *An Ocean of Air* by Gabrielle Walker (Bloomsbury).

11. This 'solid body rotation', as it is called, is a solution to the equations of motion but quite an unphysical one. In particular, any forcing such as the radiation coming in from the Sun would soon make the atmosphere unstable, and weather patterns would develop.

12. In this, the early scientists were completely correct, as we shall see when we attempt to understand the mid-latitude westerlies in Chapter 8.

13. Napier Shaw (1920), 'Pioneers in the science of weather', *Quarterly Journal of the Royal Meteorological Society*. Shaw was a distinguished meteorologist, who worked as Director of the Meteorological Office, the first Professor of Meteorology at Imperial College, London, and as the President of the Royal Meteorological Society.

14. More information on the history of the Hadley cell can be found in: 1) Persson, A. (2006). 'Hadley's principle: understanding and misunderstanding the trade winds'. *History of Meteorology*, 3, 17–42. 2) Burstyn, H. L. (1966). 'Early explanations of the role of the Earth's rotation in the circulation of the atmosphere and the ocean'. *Isis*, 57(2), 167–187.

Chapter 3

15. More detail on William Ferrel's incredible story is given by Gabriele Walker in her book *An Ocean of Air*. While his Nashville paper makes interesting reading, it does not stand the test of time as much as his truly classic work: 'The influence of the Earth's rotation upon the relative motion of bodies near its surface' (*The Astronomical Journal* 1858).

16. Here, we will simply call this the pressure force, but it is more formally termed the *pressure gradient force*. We need a spatial gradient in pressure to impart a force on the flow, acting from regions of high pressure towards regions of low pressure.

17. The *almost* is important in this sentence, as you might expect. Acceleration isn't just about speeding up or slowing down, it is also involved when something turns. For something to undergo circular motion, as the air around our cyclone is, there must be a net acceleration acting towards the centre of the circle. The next simplest approximation is called *gradient wind balance*, and in this approximation the pressure force and Coriolis add up to give a net force inwards, which is the centripetal force maintaining circular motion around the cyclone. For more detail on the development of the equations of motion for the atmosphere, and approximations to them, see introductory textbooks on meteorology, such as 1) Hoskins & James (2014). *Fluid Dynamics*

of the Mid-Latitude Atmosphere. John Wiley & Sons. 2) Andrews, D. G. (2010). *An Introduction to Atmospheric Physics.* Cambridge University Press.

Chapter 4

18. In 2019 the MOSAiC project will attempt this again, with the German icebreaker *Polarstern* frozen into the ice and used as a floating laboratory from which to study the Arctic as the ship drifts across the basin: https://www.mosaic-expedition.org.

19. For more detail on Ekman, see the excellent introductory textbook of Marshall and Plumb (*Atmosphere, Ocean and Climate Dynamics,* Academic Press). The net effect of wind blowing over the ocean is to move the seawater in the upper 50 m or so precisely 90° to the right of the wind direction (in the Northern Hemisphere). However, the effect varies with depth in what we call an *Ekman spiral.* Near the surface, the water will be flowing just a bit to the right of the wind, as Nansen observed the icebergs to move. A few tens of metres down, near the base of the *Ekman layer* the current is weak but actually directed backwards, against the wind. Only when averaged over this layer is the current directed at exactly 90 degrees.

20. See, for example, the book *Rough Crossings* by Simon Schama (Vintage).

21. The Hadley cell, however, is curious in this regard. Although driven by the ascent of the warmest air in the ITCZ, its horizontal flow of heat actually works in the wrong way. The air moving towards the equator near the surface is relatively warm, while the colder, upper level air is moving towards the pole. The Hadley cell therefore clearly transports heat in the wrong direction. The paradox is only resolved by considering the transport of gravitational potential energy: the upper level air, having been lifted high off the surface, has much more potential energy than the near surface air, and it is through the export of energy in this form to higher latitudes that the Hadley cell fulfils its mission. The working of the Hadley cell reveals an intimate secret of the atmosphere: while it does achieve its mission of transporting energy polewards, it actually does so very inefficiently.

22. Many aspects of how climate has influenced the human history of Africa are described in John Reader's book *Africa* (Penguin).

23. These disturbances are termed African Easterly Waves. They propagate west along the low-level jet, out over the tropical Atlantic Ocean. There they can provide favourable conditions for the growth of tropical storms, often referred to as Cape Verde hurricanes. Many of the strongest storms which affect the Caribbean begin life as Cape Verde hurricanes, with the help of African Easterly Waves.

24. See, for example, O'Reilly et al (2017), 'The dynamical influence of the Atlantic Multidecadal Oscillation on continental climate', *Journal of Climate.*

25. The full story of Sahel rainfall has many extra complications. As well as the African Easterly Jet there are several other important features wrapped up in what we know as the West African Monsoon. The Indian and even the distant Pacific oceans seem to have some effect, as well as the Atlantic. For a general review of the topic, see Nicholson, S. E. (2013). 'The West African Sahel: A review of recent studies on the rainfall regime and its interannual variability'. *ISRN Meteorology* 2013.

Chapter 5

26. Moore (2004): 'Mount Everest snow plume: A case study'. *Geophysical Research Letters*.
27. The change in jet speed is actually enough to cause measurable changes in the Earth's rotation. It turns out that the Southern Hemisphere jets change much less through the year than their northern counterparts. This is because the surface is more dominated by ocean which keeps its heat better, and so the temperature contrast between latitudes varies little between seasons. To a first approximation then, as the year progresses from August to December, the southern jets don't change much but the northern jets strengthen. Angular momentum has to be conserved, so with more momentum in the atmosphere there must be less in the solid earth, so the planet's rotation slows and the day gets longer, by about a millisecond.
28. For more on Rossby himself, see *Inventing Atmospheric Science: Bjerknes, Rossby, Wexler, and the Foundations of Modern Meteorology*, James Rodger Fleming (2016), MIT Press. Also Phillips, N. A. (1998). 'Carl-Gustaf Rossby: His times, personality, and actions'. *Bulletin of the American Meteorological Society*, 79(6), 1097–1112.
29. The fact that water flows out of the sink anticlockwise in the Northern Hemisphere and clockwise in the southern is, sadly, a hoax. Without the helpful swirl to get it going, it is simply random which way the water ends up spinning. As the water flows out of the plughole, columns of fluid are stretched and, as for the ice skater pulling their arms in, their vorticity is amplified. Hence, the direction of the vortex just depends on whatever direction happened to dominate in the small random bits of vorticity that were there initially.
30. For the more mathematically inclined, the relative vorticity is simply the curl of the velocity vector. From a scale analysis of the equations of motion, it is apparent that the only component of the vorticity vector which really matters, at least for Rossby waves, is the local vertical component (i.e. in the direction pointing away from the centre of the Earth).
31. Several scientific papers provide more analysis of the events of summer 2010, for example: Matsueda (2011), 'Predictability of Euro-Russian blocking in summer of 2010', *Geophysical Research Letters*. Galarneau Jr et al. (2012),

'A multiscale analysis of the extreme weather events over western Russia and northern Pakistan during July 2010', *Monthly Weather Review*. Martius et al (2013), 'The role of upper-level dynamics and surface processes for the Pakistan flood of July 2010', *Quarterly Journal of the Royal Meteorological Society*. Webster et al (2011), 'Were the 2010 Pakistan floods predictable?' *Geophysical Research Letters*.

Chapter 6

32. This story is recalled by one of the forecasters, William Plumley in his article 'Winds over Japan'. *Bulletin of the American Meteorological Society* 1994. Plumley notes that after the mission failed, the officer had the good grace to apologize, saying 'we measured the winds; they were 170 knots'.

33. Information on Second World War encounters with the jet stream was taken from: Berson, F. A. (1991), 'Clouds on the horizon: Reminiscences of an international meteorologist', *Bulletin of the American Meteorological Society*; Lewis, J. M. (2003), Ooishi's Observation: Viewed in the context of Jet Stream discovery, Bulletin of the American Meteorological Society; Phillips, N. A. (1998), 'Carl-Gustaf Rossby: His times, personality, and actions', *Bulletin of the American Meteorological Society*; *The Dance of Air and Sea* by Arnold H. Taylor, 2011, Oxford University Press.

34. Tamura, S. T. (1905). Mt. Tsukuba Meteorological Observatory, Founded by HIH Prince Yamashina'. *Science*, 22, 122–124.

35. Ooishi's story was carefully researched for an English-, not Esperanto-speaking audience by Lewis, J. M. (2003). Ooishi's Observation: Viewed in the context of Jet Stream discovery. *Bulletin of the American Meteorological Society* 84(3), 357–369.

36. Some examples of this are discussed by Lyall Watson in his popular book *Heaven's Breath*.

37. Meteorologists also often call these *extratropical cyclones*. 'Extratropics' is a useful term that simply refers to everywhere on the globe outside of the tropical belt, in the same way that 'extraterrestrial' refers to anything not from Earth. For simplicity, though, we will use 'mid-latitude' here.

38. A similar observation was made later by Benjamin Franklin, the polymath who flew a kite into a storm in 1750 in order to prove that lightning is made of electricity, miraculously survived and went on to be one of the Founding Fathers of the United States on independence. On 21 October 1743, Franklin was annoyed to miss an expected lunar eclipse in Philadelphia because of the clouds accompanying a storm. He later learnt that the eclipse was seen perfectly from Boston, where the storm only arrived the next day. By contacting people in between the two cities, Franklin deduced that this was a single storm that had worked its way east along the North American coast, even though the gales had been blowing in the opposite way, to the west.

(From this we can infer that the cyclone centre must have been tracking to the south of both cities in this case, as the easterly winds will be on the north side of the storm.)

39. Another contender for this honour is Admiral Robert Fitzroy who pioneered the first scientific weather forecasts, years after circling the globe with the young Charles Darwin. Fitzroy drew cyclones as great swirls of air, in beautiful images which look remarkably realistic, although without the level of detail in Bjerknes' pictures. He developed his forecasts using a network of observations and published them in daily newspapers. Fitzroy's forecasts saved many lives at sea but he was ridiculed by the scientific establishment and tragically took his own life.

40. For more details and images of mid-latitude cyclones, see the *Oxford Very Short Introduction to Weather*, by Storm Dunlop, or any basic meteorology textbook. Real cyclones can be examined online at the STORMS Extratropical Cyclone Atlas at the University of Reading: http://www.met.rdg.ac.uk/~storms/.

41. For more on the importance of the Norwegian cyclone model and its place in the history of meteorology, see *The Life Cycles of Extratropical Cyclones* edited by Shapiro and Gronas, *American Met. Soc.*, 1999.

42. Ironically, the German/Austrian theories of scientists such as Felix Exner had considered upper level flow crucial for cyclone growth, but these had been displaced by the Bergen model which offered more immediate practical forecasting application at the time. See, e.g. *Inventing Atmospheric Science* by James Roger Fleming.

43. A further complexity emerges here concerning the interplay between the winds and the temperatures. It turns out that a warm anomaly at the ground behaves like an cyclonic vortex. So to make a cyclone high up in the atmosphere we have to move air from a high latitude towards the equator, but near the ground it is the opposite: we move warm air towards the pole, preferably across a strong temperature gradient like that provided by the Kuroshio.

44. It was developed independently by the American scientist Jule Charney and the Englishman Eric Eady, as discussed in Chapter 8.

45. Cartwright & Nakamura (2009). 'What kind of a wave is Hokusai's Great wave off Kanagawa?' Notes and Records of the Royal Society.

46. There is a problem here, of course, in that the whole thing is going the wrong way. It would have to be a Southern Hemisphere cyclone. . . .

Chapter 7

47. Data from the Aerological Observatory at Tsukuba can be browsed here: http://www.jma-net.go.jp/kousou/index-e.html, and another excellent website (in Japanese) permits a detailed exploration of all recent extreme winter storms over Japan: http://fujin.geo.kyushu-u.ac.jp/meteorol_bomb/.

48. If the water is deep and still enough, the energy should be moving half as fast as the wave peaks and troughs. See an introductory book on fluid dynamics, such as *Elementary Fluid Dynamics* by D. J. Acheson (Oxford University Press).

49. This process of dispersion in Rossby wave packets is commonly known as downstream development. The classic paper demonstrating this from atmospheric data is Chang (1993), 'Downstream development of baroclinic waves as inferred from regression analysis', *Journal of the Atmospheric Sciences*.

50. Details from the NOAA Global Hazards Report. These useful reports are available online, for example in this case: https://www.ncdc.noaa.gov/sotc/hazards/200211.

51. For more of the story of the Friendly Floatees, see *Moby Duck* by Donovan Hohn (Union Books).

52. Ebbesmeyer et al (2007). 'Tub toys orbit the Pacific Subarctic gyre'. *Eos, Transactions American Geophysical Union*, 88(1), 1–4.

53. Hoskins & Valdes (1990). 'On the existence of storm-tracks'. *Journal of the Atmospheric Science*. A review of the role of such western boundary currents in climate is given by Kwon et al (2010). 'Role of the Gulf Stream and Kuroshio-Oyashio systems in large-scale atmosphere-ocean interaction: A review'. *Journal of Climate*.

54. See, for example, the *Very Short Introduction to Climate Change*, by Mark Maslin, Oxford University Press.

Chapter 8

55. Another great figure of twentieth century meteorology, Victor Starr, was also involved. Starr is best known for his work demonstrating how the shape and structure of weather systems helped them to feed momentum into the jet, as described later in this chapter.

56. Raymond Hide at Cambridge performed similar experiments a little while later, attempting to model the flow of molten rock in the Earth's core. Part of the wonder of fluid dynamics is that it can describe this equally as well as the motion of the atmosphere and the oceans.

57. Fultz soon switched to a different type of apparatus. The hemispheric shell suffered from the fundamental problem that the gravity force pulled the fluid down towards the South Pole, not the centre of the model planet. Fultz took a step backward in complexity, giving up the spherical geometry for a flat, rotating dishpan of water. These are his most famous experiments, versions of which are still repeated in teaching labs around the world. For a technical review of similar laboratory experiments used in geophysical fluid dynamics, see Read et al, 'General Circulation of Planetary Atmospheres: Insights from Rotating Annulus and Related Experiments', Chapter 1 of *Modeling Atmospheric and Oceanic Flows: Insights from Laboratory Experiments and Numerical*

Simulations; John Wiley & Sons, 2015; pp. 7–44. The excellent introductory textbook by Marshall and Plumb has online links to several interesting lab experiments (*Atmosphere, Ocean and Climate Dynamics: An Introductory Text*, Academic Press, 2007).

58. 'On the General Circulation of the Atmosphere in Middle Latitudes'; *The Bulletin of the American Meteorological Society*, 1947, Vol 28, no. 6, pp. 255–280.

59. This story was reported in Machta (1992), *Bulletin of the American Meteorological Society*, 73, 1797–1806. A similar case occurred in May 1962, when dairies in Kansas and Missouri detected unusually high levels of radiation in routine milk tests. This time the culprit was the US itself, which had conducted a nuclear test near Christmas Island to the south of Hawaii. Unfortunately, an unusual wind pattern had carried some of the contamination into the Pacific jet stream, and the rest is history (recorded in Reiter's book *Jet Streams: How do they Affect our Weather?*).

60. See, for example, the mini-biography of Charney in 'Storm Watchers', by John D Cox, Wiley.

61. In the words of Prof. Geoff Vallis (University of Exeter): '[Eady] is one of the very few scientists of whom it can be fairly said that we would have liked for him to have published more'.

62. For a bit more on Eady, see his obituary in the *Quarterly Journal of the Royal Meteorological Society*, and also his entry in the National Dictionary of Biography; http://www.oxforddnb.com/view/article/37381. His classic paper is 'Long waves and cyclone waves', published in 1949 in *Tellus*, the journal that Rossby initiated in Stockholm.

63. The first of several classic papers on these was Simmons & Hoskins (1978): 'The life cycles of some nonlinear baroclinic waves', *Journal of the Atmospheric Sciences*.

64. This was also in agreement with observations. Meteorologists such as Starr and Palmen had first measured momentum fluxes such as these in early upper air data in the 1950s. A simple understanding of the momentum fluxes goes as follows: The winds associated with the eddy are tilted to either blow to the northeast or to the southwest. The former of these can be thought of as a wind to the north moving a little bit of westerly momentum north. Similarly, the latter can be thought of as a wind to the south moving a little bit of easterly momentum south. In both cases there is an effective northward transport of westerly momentum, which is what we are after.

65. Lorenz, E. N. (1955). 'Available potential energy and the maintenance of the general circulation'. *Tellus*, 7(2), 157–167. Lorenz would go on to even greater fame as the inventor of chaos theory.

66. For another example of the use of the momentum budget, consider the effects of surface drag, in which the Earth's surface acts to slow down the wind blowing over it. In the mid-latitudes this wind is westerly, so the drag acts to reduce the westerly momentum of the atmosphere. In the tropics,

however, the drag acts on the easterly trade winds, hence reducing the easterly momentum of the atmosphere, or equivalently, increasing the westerly momentum. So in terms of westerly momentum, we see that this is generated by drag in the tropics, and destroyed by drag in the mid-latitudes. Hence, the atmospheric circulation has to move westerly momentum from the tropics to the mid-latitudes, and both the Hadley Cell and the storm tracks help to do this job.

67. Reiter, E. R. (1963), *Jet-stream Meteorology*, University of Chicago Press. Reiter, E. R. (1967), *Jet streams. How do they affect our weather?* Doubleday Anchor.

68. A similar effect could be achieved by speeding up the planet's rotation. Classic experiments of this nature were performed by Williams, G. P. (1978). 'Planetary circulations: 1. Barotropic representation of Jovian and terrestrial turbulence'. *Journal of the Atmospheric Sciences*, 35(8), 1399–1426.

69. This is a common trick used to mimic atmospheric flow in fluids labs. Think of the water in the tank as made up of columns of fluid which each like to stay as a vertical column (it behaves like this due to the rotation). As a column moves up the sloping bottom, for example, it is squashed and hence its vorticity is reduced. This is because it spins more slowly, like the ice skater spreading out their arms. Hence, if the bottom slopes upward towards the middle of the tank, water moving towards the middle develops an anticyclonic motion, just as in air moving poleward in the real atmosphere. For more information on this experiment, see Read et al (2015), 'An experimental study of multiple zonal jet formation in rotating, thermally driven convective flows on a topographic beta-plane', *Physics of Fluids* 27, 085111.

70. Or at least, the jet spacing is closely related to the Rhines scale. Dritschel & McIntyre (2008), *Journal of the Atmospheric Sciences* discuss the relationship between the two for their model experiments, twenty-first century versions of Rhines'.

71. This is characteristic of two-dimensional turbulence, which is the type that can be applied to the large-scale flow in both the atmosphere and ocean. This contrasts strongly with the more familiar three-dimensional turbulence, such as the waves crashing on the beach, where the vortices get smaller over time, creating millions of tiny swirls and filaments.

72. As an aside, the eddies are also responsible for the existence of the Ferrel cell: the mid-latitude version of the Hadley cell which was first predicted by the unassuming ex-farmhand, William Ferrel. This cell is a staple of geography textbooks and depicts air rising upward at around 60° of latitude, moving equatorward and then sinking. The obvious puzzle here is why the colder, subpolar air rises and the warmer, subtropical air sinks? In thermodynamics language this is a thermally indirect system, in that energy has to be pumped in to the system to achieve this feat, and in this case what does the pumping is all the eddies. The Ferrel cell is a real feature of the atmosphere but conceptually it is not that important. The winds associated with it are much weaker

than those of the eddies themselves, so it is very much a weak background against which the all-important turbulent drama plays out. If you could tag a parcel of air and follow it around the Hadley cell, chances are that it would do just what the textbook cartoons say it should. Do the same in the Ferrel cell, however, and it would disregard the cartoon entirely and go wherever the weather patterns of the day will take it.

73. Rhines' classic paper on jets was published in 1975 (Waves and turbulence on a beta-plane. *Journal of Fluid Mechanics*, 69(3), 417–443). For a more general discussion of jet formation mechanisms in a mathematically based textbook, see *Atmospheric and Oceanic Fluid Dynamics* by G. K. Vallis, Cambridge University Press.

74. Specifically, it is the *stratification*, i.e. the rate of change of density in the vertical, which determines the horizontal length scale. Density in the ocean changes much more gradually with height than that in the atmosphere, so the ocean stratification is weaker. Flow structures often form at a length scale called the Rossby deformation radius (yes, him again), which is the scale where rotation and stratification are equally important. As the stratification is weaker in the ocean, the effect of rotation can be weaker as well, so the deformation radius can be smaller.

Chapter 9

75. This quote has been attributed to Gene Rasmusson, who quickly changed his tune and went on to do fundamental work on ENSO himself; see *El Niño* by J. Madeleine Nash, Warner Books.

76. An accessible overview is given by Walker, Gilbert (1928). 'World weather'. *Quarterly Journal of the Royal Meteorological Society*. 54 (226): 79–87.

77. Oceanic fluid dynamics is important here, as the spreading of heat is accomplished by dynamical waves known as Kelvin waves. For some more detail on the dynamics of ENSO, see Marshall & Plumb (2016): *Atmosphere, Ocean and Climate Dynamics: An Introductory Text*, Academic Press. There is excellent educational material, as well as information on current ENSO conditions and forecasts, on the NOAA CPC website: http://www.cpc.ncep.noaa.gov/products/precip/CWlink/MJO/enso.shtml.

78. A key paper here is Lu et al (2008). 'Response of the zonal mean atmospheric circulation to El Niño versus global warming'. *Journal of Climate*.

79. For a popular account of ENSO impacts, see the book *El Niño* by J. Madeleine Nash. An ENSO link to the Russian heatwave was suggested by Schneidereit et al (2012), in *Monthly Weather Review*.

80. *El Niño* by J. Madeleine Nash.

81. For more on the California drought, see Swain (2015), A tale of two California droughts: Lessons amidst record warmth and dryness in a region of complex physical and human geography, *Geophysical Research Letters*. For the meteo-

rological drivers in particular see Seager et al (2015), 'Causes of the 2011–14 California drought', *Journal of Climate*.

82. A review of scientific work on the slowdown was performed by Xie & Kosaka (2017), 'What caused the global surface warming hiatus of 1998–2013?' *Current Climate Change Reports*.

83. Kumar et al (2006), 'Unraveling the mystery of Indian monsoon failure during El Niño', *Science*.

84. For more on the 2015 event, and the predictions of it, see L'Heureux et al (2017), 'Observing and Predicting the 2015/16 El Niño'. *Bulletin of the American Meteorological Society*.

Chapter 10

85. See the project website hosted at the University of Washington (http://olympex.atmos.washington.edu) or the overview paper: Houze Jr, R. A., et al (2017), 'The Olympic Mountains Experiment (OLYMPEX)', *Bulletin of the American Meteorological Society*.

86. This is a slightly unfair comparison actually. While the Himalayan peaks themselves don't impact the jet much, the high plateaus of Tibet and Mongolia further north have a very large effect on the atmospheric circulation in general. These impact the low-level westerly flow further north and hence set up downstream Rossby waves. It is partly these waves which keep the jet oriented so straight from west to east across the Pacific Ocean. See Saulière et al (2012), 'Further investigation of the impact of idealized continents and SST distributions on the Northern Hemisphere storm tracks', *Journal of the Atmospheric Sciences*; and White et al (2017), 'Mongolian Mountains Matter Most: Impacts of the Latitude and Height of Asian Orography on Pacific Wintertime Atmospheric Circulation', *Journal of Climate*.

87. The classic review paper on stationary waves is Held et al (2002), 'Northern winter stationary waves: Theory and modeling', *Journal of Climate*. As well as orography, the observed stationary wave pattern is strongly shaped by the distribution of heating within the atmosphere.

88. *A Very Short Introduction to Climate*, by Mark Maslin (Oxford University Press), has a concise summary of past Earth climates.

89. Caballero, R., & Huber, M. (2010). 'Spontaneous transition to superrotation in warm climates simulated by CAM3'. *Geophysical Research Letters*.

90. See Merz et al (2015). 'North Atlantic eddy-driven jet in interglacial and glacial winter climates'. *Journal of Climate*, and references therein.

91. The dynamics of this is a little harder, and involves the *ageostropic* wind, i.e. that part of the wind field which is not in geostrophic balance with the pressure field. As the wind accelerates into the jet, an ageostrophic wind is generated which blows to the left of the acceleration, i.e. to the north in this

case. You can think of this occurring because the wind has not accelerated quickly enough, so that Coriolis is not large enough to balance the pressure gradient, which therefore turns the wind towards the lower pressure on the left. This sets up a systematic drift northward, from the right entrance region toward the left entrance region. This in turn generates divergence in the right entrance and convergence in the left entrance. The column of air underneath the right entrance region is therefore prone to ascent which enhances storm growth. The reverse happens in the jet exit regions, where the ageostrophic flow is toward the south.

92. There are a few slightly different theories like this, but the one which has received the most attention is Francis & Vavrus (2012), 'Evidence linking Arctic amplification to extreme weather in mid-latitudes', *Geophysical Research Letters* 39. Another interesting theory is that of Petoukhov et al (2013), 'Quasiresonant amplification of planetary waves and recent Northern Hemisphere weather extremes', *Proceedings of the National Academy of Sciences*, though this also suffers from the short period considered and the lack of evidence of causality.

93. Wallace et al (2014), 'Global warming and winter weather' *Science*, 343, p. 729.

94. A good review of the issues, distinguishing three different important questions, is provided by Barnes & Screen (2015), 'The impact of Arctic warming on the midlatitude jet-stream: Can it? Has it?' Will it? *Wiley Interdisciplinary Reviews: Climate Change*. (Spoiler: the three answers are: 1) yes; 2) no, at least not 'significantly'; 3) yes, but so will lots of other things. . . .)

95. See for example, Watson et al (2016), 'The role of the tropical West Pacific in the extreme Northern Hemisphere winter of 2013/2014,' *Journal of Geophysical Research: Atmospheres*.

Chapter 11

96. This quote, along with other details of this story, is taken from Glaisher's book, *Travels in the Air*. See also *The Weather Experiment*, by Peter Moore (Chatto & Windus). For a historical account of Glaisher's life and work, see *Storm Watchers* by John D Cox, (Wiley).

97. Though Glaisher, at fifty-seven, was already a Fellow of the Royal Society, a pioneer of weather reporting and a stalwart of the British science community. He went on to lead the Royal Meteorological Society as its President.

98. For a scientific history of the Gulf Stream, see *The Gulf Stream* by Stommel (1950), *Scientic American*, vol 4.

99. More on the history of Franklin's map is given, for example, by Bache (1936), 'Where Is Franklin's First Chart of the Gulf Stream?' *Proceedings of the American Philosophical Society*.

100. In the words of John D Cox, in his book *Storm Watchers*, Maury's book 'often has been referred to as the founding textbook on the subject of oceanography, especially by people who have not read it'.
101. Seager et al (2002), 'Is the Gulf Stream responsible for Europe's mild winters?' *Quarterly Journal of the Royal Meteorological Society*.
102. Brayshaw et al (2009 & 2011) 'The basic ingredients of the North Atlantic storm track, Parts 1 & 2. *Journal of the Atmospheric Sciences*.

Chapter 12

103. Li & Wettstein (2012). 'Thermally driven and eddy-driven jet variability in reanalysis'. *Journal of Climate*.
104. This situation is in fact not unique. There is also an example in the Southern Hemisphere winter, when the jet over the South Pacific is often split into two parallel jets. This is also a local effect caused by east-west asymmetries, in this case the distribution of warmth in the tropics and the gradients of sea surface temperature in the mid-latitudes. Inatsu & Hoskins (2004), 'The zonal asymmetry of the Southern Hemisphere winter storm track', *Journal of Climate*.
105. *Climate, History and the Modern World* by H. H. Lamb (Routledge).
106. Brohan et al (2010). 'Arctic marine climate of the early nineteenth century'. *Climate of the Past*.
107. See, e.g. Rigor & Wallace (2004), 'Variations in the age of Arctic sea-ice and summer sea-ice extent', *Geophysical Research Letters*. For lots of online information about sea ice, including the latest news, go to the US National Snow and Ice Data Center at https://nsidc.org.
108. There have been suggestions that the King's Mirror, an ancient Norwegian text from around 1250AD, reveals some understanding of the nature of North Atlantic climate variability. Given the complexity of this variability it is doubtful the early Norwegians had a very sophisticated understanding of this, though there are statements such as: '. . . all the regions that lie near get severe weather from this ice, inasmuch as all the storms that the glacier drives away from itself come upon others with keen blasts'. This seems to imply an understanding that when one region is spared by the storms, they have simply moved to another region, which is ultimately not a bad description of storm track variability. For more discussion, see Haine, T. (2008). 'What did the Viking discoverers of America know of the North Atlantic Environment?' *Weather*, 63(3), 60–65.
109. See Hans Volkert's piece on Exner in the New Dictionary of Scientific Biography (Charles Scribner's Sons). A similarly pioneering attempt at weather forecasting was made by Englishman Lewis Fry Richardson just a little later. Richardson famously completed his calculations by hand while working

as an ambulance driver in the trenches of the First World War. See, for example, *Storm Watchers* by John D. Cox.

110. Hurrell, & Deser (2015), 'Northern Hemisphere climate variability during winter: Looking back on the work of Felix Exner', *Meteorologische Zeitschrift*.

111. For more on the history of research on the NAO, see Stephenson et al (2003): 'The history of scientific research on the North Atlantic Oscillation', in *The North Atlantic Oscillation: Climatic Significance and Environmental Impact*, Geophysical Monograph, American Geophysical Union.

112. More sophisticated NAO indices use data from all across the North Atlantic, incorporated in a statistical method such as principal component analysis, but these similarly reflect average conditions over the region. Hurrell & Deser (2010). 'North Atlantic climate variability: the role of the North Atlantic Oscillation'. *Journal of Marine Systems*.

113. Woollings et al (2010). 'Variability of the North Atlantic eddy-driven jet stream'. *Quarterly Journal of the Royal Meteorological Society*.

114. For these, and other NAO-related impacts of the jet, see the NAO book: *The North Atlantic Oscillation: Climatic Significance and Environmental Impact*, Edited by Hurrell et al, AGU Geophysical Monograph Series.

115. Catchpole & Faurer (1983). 'Summer sea ice severity in Hudson Strait', 1751–1870, *Climatic Change*.

Chapter 13

116. A thorough review of the effects of volcanos on climate is given by Robock (2000), 'Volcanic eruptions and climate'. *Reviews of Geophysics*.

117. For more on the societal and historical consequences of the year without summer, see the popular book '1816' by Klingaman and Klingaman (St Martin's Press). A scientific review of the impacts of Tambora is given by Raible et al (2016), 'Tambora 1815 as a test case for high impact volcanic eruptions: Earth system effects'. *Wiley Interdisciplinary Reviews: Climate Change*.

118. These dramatic vortex breakdown events were revealed in early satellite data by McIntyre & Palmer (1983), 'Breaking planetary waves in the stratosphere'. *Nature*.

119. For a short review of the coupling between the stratosphere and the troposphere, see Kidston et al (2015), 'Stratospheric influence on tropospheric jet streams, storm tracks and surface weather', *Nature Geoscience*.

120. A collection of work on Krakatoa was compiled by the Royal Society a few years after the event: Symons et al (1888), *The Eruption of Krakatoa, and Subsequent Phenomena*. A recent perspective including the link to the QBO is given by Hamilton (2012), Sereno Bishop, Rollo Russell, Bishop's Ring and the Discovery of the 'Krakatoa Easterlies', *Atmosphere-Ocean*.

121. An overview of possible drivers of NAO variability is given by Smith et al (2016), 'Seasonal to decadal prediction of the winter North Atlantic

Oscillation: emerging capability and future prospects'. *Quarterly Journal of the Royal Meteorological Society*.

122. Fereday et al (2012), 'Seasonal forecasts of northern hemisphere winter 2009/10', *Environmental Research Letters*.

123. Greatbatch et al (2015), Tropical origin of the severe European winter of 1962/1963, *Quarterly Journal of the Royal Meteorological Society*.

124. Knight et al (2017), 'Global meteorological influences on the record UK rainfall of winter 2013–14', *Environmental Research Letters*.

125. Maidens et al (2013), 'The influence of surface forcings on prediction of the North Atlantic Oscillation regime of winter 2010/11', *Monthly Weather Review*.

126. Sutton and Dong (2012), 'Atlantic Ocean influence on a shift in European climate in the 1990s,' *Nature Geoscience*.

127. The original hindcasts were reported in Scaife et al (2014), 'Skillful long-range prediction of European and North American winters', *Geophysical Research Letters*, with the more extended period and the inclusion of the second winter in Dunstone et al (2016): 'Skilful predictions of the winter North Atlantic Oscillation one year ahead', *Nature Geoscience*.

128. Lorenz did later adopt the more picturesque butterfly, as in one of his famous talks: Does the flap of a butterfly's wings in Brazil set off a tornado in Texas? Lorenz's popular book *The Essence of Chaos* recounts some of the history and is a useful introduction to the topic.

129. Scaife & Smith (2018), 'A signal-to-noise paradox in climate science', *NPJ Climate and Atmospheric Science*.

Chapter 14

130. Statistics here come from the Global Wind Energy Council in their Global Wind Report 2017.

131. *Heaven's Breath* by Lyall Watson (Coronet Books).

132. See, for example, Marvel et al (2013), 'Geophysical limits to global wind power', *Nature Climate Change*.

133. The mechanism here is similar to that described in Chapter 10 for the jet entrance, just with different directions. At the end of the jet, the wind is decelerating, or equivalently feels an acceleration towards the west. The ageostrophic wind blows to the left of this acceleration, i.e. from north to south, and hence leads to divergence to the north (in the left exit) and convergence to the south (in the right exit). This is all strongest in the upper reaches of the troposphere, at Grantley's level where the jet is strongest, and the rest of the air column has to adjust to this. The end effect is ascent under the left exit of the jet but descent under the right exit.

134. Fink et al (2009), 'The European storm Kyrill in January 2007: synoptic evolution, meteorological impacts and some considerations with respect to climate change', *Natural Hazards and Earth System Sciences*.

135. Lothar is also notable for an unusual, 'bottom-up' development, in which diabatic processes, such as those associated with clouds and precipitation, are thought to have aided the rapid growth of the storm; Wernli et al (2002), 'Dynamical aspects of the life cycle of the winter storm "Lothar" (24–26 December 1999)', *Quarterly Journal of the Royal Meteorological Society*.

136. In several cases a distinct dynamical setup occurs, with the jet pinned in the middle of two Rossby wave breaking events, one to the north and one to the south. These help to keep the jet strong, and at the same time stuck in the same place. See, e.g. Pinto et al (2014), 'Large-scale dynamics associated with clustering of extratropical cyclones affecting Western Europe', *Journal of Geophysical Research: Atmospheres*.

137. This is perhaps a little unfair on the north and south winds. By bringing either cold, dry air from the north, or warm, moist air from the south, these winds can have a very important influence on the amount of rainfall in any given location. For example, see Simpson et al (2016), 'Causes of change in Northern Hemisphere winter meridional winds and regional hydroclimate', *Nature Climate Change*.

138. *Heaven's Breath*, by Lyall Watson (Coronet Books).

139. Nakamura, N., & Huang, C. S. (2018). 'Atmospheric blocking as a traffic jam in the jet stream'. *Science*.

140. *Storm Surge*, by Adam Sobel (Harper Collins).

141. Petersen (2010). 'A short meteorological overview of the Eyjafjallajökull eruption 14 April–23 May 2010'. *Weather*.

142. Woods et al (2013). 'Large-scale circulation associated with moisture intrusions into the Arctic during winter'. *Geophysical Research Letters*.

Chapter 15

143. The regular reports from the Intergovernmental Panel on Climate Change remain the authoritative source of information on the science of climate change (www.ipcc.ch), but much shorter and more readable coverage is provided by Mark Maslin in *Climate Change: A Very Short Introduction* (Oxford University Press).

144. We have referred to computer models or simulations of climate at various points in the book. These grew out of weather forecast models, stepping forwards in time following Newton's laws, but now include a myriad of additional processes which don't matter so much on day-to-day weather timescales, from ocean circulation and sea ice to vegetation and chemistry. A nice introduction to climate models is available at https://www.carbonbrief.org/qa-how-do-climate-models-work.

145. Though very drastic action will be required if we want to limit the total warming to 1.5°C above pre-industrial, as was suggested by the politicians at the United Nations Climate Change Conference (COP21) in Paris in 2015.

146. This is really just a conceptual picture of the radiative changes. The simplest model that gets the numbers roughly right adjusts to the change in composition by changing the effective level of emission of radiation to space, and the whole vertical temperature structure of the atmosphere changes along with that. See Archer (2012): *Global Warming: Understanding the Forecast* (Wiley).

147. Some simple and concise discussion of this and other effects on the structure of the atmosphere are given in Vallis et al (2015), 'Response of the large-scale structure of the atmosphere to global warming', *Quarterly Journal of the Royal Meteorological Society*.

148. Held & Soden (2006). 'Robust responses of the hydrological cycle to global warming'. *Journal of Climate*.

149. Joshi et al (2008). 'Mechanisms for the land/sea warming contrast exhibited by simulations of climate change'. *Climate Dynamics*.

150. A nice commentary on this contrast is given by Shepherd (2014), 'Atmospheric circulation as a source of uncertainty in climate change projections', *Nature Geoscience*.

151. At least in the Northern Hemisphere; many models do predict an increase in intense storms in the Southern Hemisphere, where polar amplification is much weaker. Also, this statement refers specifically to mid-latitude cyclones. Their tropical counterparts, including hurricanes in the Atlantic and typhoons in the Pacific, are expected to get more intense as the planet warms. This is because they derive their energy from the difference in temperature between the bottom of the storm (at the sea surface where they evaporate water) and the top (in the lowermost stratosphere). This temperature difference is expected to increase as the surface warms but the stratosphere cools. See *Storm Surge* by Adam Sobel for more detail.

152. Hall et al (1994). 'Storm tracks in a high-resolution GCM with doubled carbon dioxide'. *Quarterly Journal of the Royal Meteorological Society*.

153. Note that this discussion is specific to the Northern Hemisphere. The southern story is a little different as polar amplification there is much weaker, due largely to much of the ice being on land. Hence, there is not so much of a competition between upper and lower levels in the south.

154. It is not all about the decrease in albedo, or reflectivity, of the surface as snow and ice melt, as often suggested. This, after all, should have no effect in the darkness of the Arctic winter, when the strongest amplification of warming is in fact predicted.

155. For recent overviews of the literature on future storm track changes, see Vallis et al (2015), Response of the large-scale structure of the atmosphere to global warming, Quarterly Journal of the Royal Meteorological Society; and Shaw et al (2016), 'Storm track processes and the opposing influences of climate change', *Nature Geoscience*. The latter paper also discusses still other

effects which might influence the storm tracks, such as changes in cloud cover.

156. Two excellent papers unpacking the details of predicted jet stream changes are Barnes & Polvani (2013; 'Response of the midlatitude jets, and of their variability, to increased greenhouse gases in the CMIP5 models', *Journal of Climate*) and Simpson et al (2014; 'A diagnosis of the seasonally and longitudinally varying midlatitude circulation response to global warming', *Journal of the Atmospheric Sciences*).

157. Woollings et al (2012), 'Response of the North Atlantic storm track to climate change shaped by ocean-atmosphere coupling'. *Nature Geoscience*. The ever-present hand of the tropics is also likely important here; see Harvey et al (2015, 'Deconstructing the climate change response of the Northern Hemisphere wintertime storm tracks, Climate Dynamics') and Ciasto et al (2016, 'North Atlantic storm-track sensitivity to projected sea surface temperature: Local versus remote influences', *Journal of Climate*).

158. Bjerknes: Half a century of change in the 'meteorological scene', *Bulletin of the American Meteorological Society*, 45 (1964), 312–315. Reprinted in History of Meteorology 6 (2014). A discussion of the role of seasonal prediction in evaluating climate models is given by Palmer et al (2008), 'Toward seamless prediction: Calibration of climate change projections using seasonal forecasts', *Bulletin of the American Meteorological Society*.

Chapter 16

159. See David Archer's (2012) book (*Global Warming: Understanding the Forecast*; Wiley) for example.

160. Thorne et al (2011). 'Tropospheric temperature trends: History of an ongoing controversy'. *Wiley Interdisciplinary Reviews: Climate Change*.

161. This pattern of ocean temperature variability is often referred to as the Pacific Decadal Oscillation, or PDO. The recent multidecadal shift to a negative PDO pattern to some extent resembles a broader, smoothed out La Niña pattern. The response of the subtropical jets is to shift poleward, i.e. in the opposite direction to during El Niño. A recent review of observed tropical widening, including the role of the PDO, is given by Staten et al (2018): 'Re-examining tropical expansion', *Nature Climate Change*. For a detailed analysis of observed changes in the jets, in particular the small poleward shift of the subtropical jet, see Manney & Hegglin (2018). 'Seasonal and Regional Variations of Long-Term Changes in Upper-Tropospheric Jets from Reanalyses'. *Journal of Climate*.

162. See, for example, Hawkins & Sutton (2012), 'Time of emergence of climate signals', *Geophysical Research Letters*.

163. This statement is based on analysis of the Northern Annular Mode (NAM), a jet-related pattern similar to the NAO, by Deser et al (2012): 'Uncertainty in

climate change projections: the role of internal variability', *Climate Dynamics*. The NAM integrates flow over a very large area. At any individual location, the clear emergence of a climate change signal in the jet is even later, in the latter half of this century (Zappa et al (2015). 'Improving climate change detection through optimal seasonal averaging: The case of the North Atlantic jet and European precipitation'. *Journal of Climate*.).

164. Woollings et al (2014). 'Twentieth century North Atlantic jet variability'. *Quarterly Journal of the Royal Meteorological Society*.
165. These fluctuations in the amount of variability seem to be linked to the decadal changes in jet speed just mentioned; Woollings et al (2018), 'Daily to decadal modulation of jet variability', *Journal of Climate*.
166. These analogies were first introduced by climate scientists James Hansen and Gerald Meehl, respectively.
167. Stott et al (2004), Human contribution to the European heatwave of 2003, *Nature*. For a scientific review of work in this area, see Stott et al (2016), 'Attribution of extreme weather and climate-related events', *Wiley Interdisciplinary Reviews: Climate Change*. A nice overview of this field is given on the CarbonBrief website: https://www.carbonbrief.org/mapped-how-climate-change-affects-extreme-weather-around-the-world.
168. As well at the Stott et al (2004) paper, see Barriopedro et al (2011), 'The hot summer of 2010: redrawing the temperature record map of Europe', *Science*.
169. Otto et al (2012). 'Reconciling two approaches to attribution of the 2010 Russian heat wave'. *Geophysical Research Letters*.
170. Some examples of different views are: Trenberth et al (2015), 'Attribution of climate extreme events', *Nature Climate Change*; Otto et al (2016), 'The attribution question', *Nature Climate Change*.
171. Cattiaux et al (2010). 'Winter 2010 in Europe: A cold extreme in a warming climate'. *Geophysical Research Letters*.
172. See, for example, Diffenbaugh et al (2015), 'Anthropogenic warming has increased drought risk in California', *Proceedings of the National Academy of Sciences*.
173. Pall et al (2011). 'Anthropogenic greenhouse gas contribution to flood risk in England and Wales in autumn 2000'. *Nature*.
174. Schaller et al (2016). 'Human influence on climate in the 2014 southern England winter floods and their impacts'. *Nature Climate Change*.

Chapter 17

175. This campaign is described and the data analysed in Angell (1959), 'A climatological analysis of two years of routine transosonde flights from Japan', *Monthly Weather Review*.
176. As an aside, this book is probably guilty of giving the impression that all of the interesting and important weather happens because of the jet stream.

This is certainly not the case, and there are many amazing stories of high-impact weather for which the jet stream is relatively unimportant, and the main actors are the systems which make up the side roads of meteorology.

177. Martius (2014). 'A Lagrangian analysis of the Northern Hemisphere sub-tropical jet'. *Journal of the Atmospheric Sciences*. It is possible to calculate your own trajectories to follow air parcels around the world, using the online HYSPLIT software provided by the US National Oceanic and Atmospheric Administration: https://www.ready.noaa.gov/HYSPLIT.php.

178. Sections on Lagrange have drawn heavily on an essay by George Sarton (1944, 'Lagrange's personality', *Proceedings of the American Philosophical Society*) and his biography in the MacTutor History of Mathematics archive by O'Connor & Robertson; University of St Andrews. While modern terminology draws a clear line between Eulerian and Lagrangian approaches, it is likely that both men also used the other's methods to some extent. See, for example, *A History and Philosophy of Fluid Mechanics* by G. A. Tokaty (Dover Publications).

FIGURE CREDITS

Fig. 1.1: Courtesy of Prof. Giles Harrison, University of Reading.

Fig. 2.2: By Edmond Halley [Public domain], via Wikimedia Commons.

Fig. 5.1: Image courtesy of the Earth Science and Remote Sensing Unit, NASA Johnson Space Center (NASA image ISS008-E-13304, http://eol.jsc.nasa.gov).

Fig. 6.1: Courtesy of the Geophysical Institute, University of Bergen.

Fig. 6.2: Bjerknes and Solberg (1922), Geofysiske Publikasjoner, vol. 3. Reproduced courtesy of Norsk Geofysisk Forening.

Fig. 6.3: Courtesy of NOAA-NASA GOES Project and NASA Earth Observatory.

Fig. 6.5: H. O. Havemeyer Collection, Bequest of Mrs. H. O. Havemeyer, 1929.

Fig. 8.1: Taken from Fultz (1950), Tellus, vol 2, p137–149. Reproduced under the terms of the Creative Commons Attribution License (http://creativecommons.org/licenses/by/4.0/).

Fig. 8.2: © Royal Meteorological Society. From Thorncroft et al (1993), Quarterly Journal of the Royal Meteorological Society, 119, 17–55.

Fig. 8.3: NASA, ESA, and A. Simon (Goddard Space Flight Center) - http://www.spacetelescope.org/images/heic1410a/.

Fig. 11.1: London Illustrated News.

Fig. 11.3: Library of Congress/Benjamin Franklin - https://www.loc.gov/resource/g9112g.ct000136/.

Fig. 17.1: Reproduced from Angell (1959), Monthly Weather Review, 87, p427–439.

All other illustrations were designed and created by Claire Delsol and Tim Woollings. The endpapers map was designed and created by Claire Delsol.

INDEX

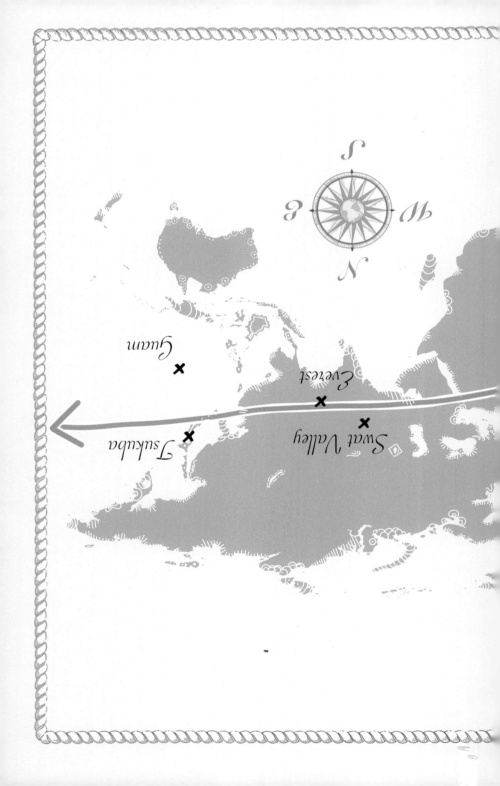